Chance,

origin and

reason of life

2023 by Roberto Guglielmo

Codice ISBN: 9798396121812

Premise

Believing in the fundamental stones of life is not easy. Such belief would seem to be automatic, as with many things that come under the senses, here and now.

I believe in something that I think or say when what I think or say is confirmed by facts. Actually, two distinct operations: what I say and what I find in the facts.

Regarding the origin of life, there are essentially two hypotheses in comparison, namely: the origin of life starting from absolute chance or by the hand of a creative intelligence. But unfortunately, the operation of establishing the fact is missing. No one can declare himself a witness at the moment of the birth of life.

Then we are free to believe or choose for one hypothesis over the other.

The book proposed by me investigates the reasons for the credibility of one hypothesis with respect to the other. Believing in an origin of life starting from absolute chance leads to conclusions of life to be lived that are profoundly different from believing in the origin of life caused by a creative intelligence.

The architecture of the book is similar to the presentation of scientific papers in general: first of all, the problem that motivates the work is posed, then the materials and methods that will be used follow, and finally the results and the final considerations or conclusions.

What has been said can be deduced from the succession of chapters.

In each chapter you can find a very short written beginning of what the chapter is about. Reason why, if the topic exposed is well known, the chapter can be skipped without losing the total meaning of the text.

I consider absolutely not to be overlooked, in order to understand the meaning of the book, the reading relating to the first three chapters, namely "Introduction", "Position of the problem" and "argumentative method". Then the other last three that is "The wonder of life", "Death. Why" and the "conclusion".

Obviously, the reading of some other specific selected chapters can always be an object of reflection, even if this could seem a mere description of what is already known,

RG

Chance, origin and reason of life

Summary

Introduction

Every dialogue, debate, exhibition, or communication arises from a problem. The evolution of living organisms presupposes a first living organism that can evolve. So, the origin of life is a problem to be addressed separately and, therefore, different from evolution issues

My friend Sergio, a philosopher by profession, and I exchanged ideas in commenting on the topic of a conference regarding natural evolution held near Piazza Navona in Rome. The stretch of road that separates Piazza Navona from Piazza di Spagna to go to the metro station is a journey that takes about twenty minutes on foot to complete. It is, apparently, enough time to compare our opinions. Sergio asserted the absence of randomness in determining the evolution of life and supported the direct intervention of God in establishing the course of natural development itself. The idea of a direction in evolution seemed to me to contrast with the theory that sees a mechanism of random mutations, with a selection of the most favorable for the organism,

Introduction

following the idea of natural selection theorized by Charles Darwin about evolution. Needless to say, twenty minutes were not enough to clarify my thought. It was clear by common sense that it was first necessary to distinguish the phenomenon of the evolution of living organisms from the problem of the birth of life in the passage leading from the world of inorganic and unstructured material towards the first living being as functionally structured for movement and reproduction. It was impossible to speak in the crowded subway section that we took at the Piazza di Spagna station. Even later, I could not clarify my standpoint, as sometimes happens to each of us, in handling concepts completely clear and familiar. In fact, in continuing my reasoning, I realized that the idea of randomness applied to birth and the evolution of life was not as clear and definable as I thought. I, therefore, proposed my friend make a subsequent meeting to illustrate better what I knew through the reminiscence of my biology studies. So, to better clarify some concepts, I put my ideas down and wrote them to better show and face the future conceptual clash. I have not renewed the meeting, but I have written this book.

I began to look for explanatory models regarding the origin and progress of life, adhering to today's scientific system of knowledge, therefore without altering the natural laws and without resorting to miraculous explanations. In adherence to these purposes, I believe that current knowledge on molecular biology can be usefully used today by researchers to allow a more precise exposition of these problems. This operation relates to evaluating the formation probability for DNA, RNA, or proteins necessary to emerge as the first and simplest living organisms. That evaluation could be accomplished through combinatorial mathematics analyses made possible by the recognized numerical composition of the subunits in the chains of proteins and DNA. An assessment that may affect the trust and credibility that everyone has to place on the various hypotheses in the field.

Of course, the concept of biological evolution rests on the fact that there must be a first living being that can reproduce itself, which then becomes subject to mutations in subsequent generations. It, therefore, seems entirely appropriate to better talk about the creation and evolution of living beings through random or non-random mutations to precede these arguments with an accurate analysis of life-origin problems.

The answer of empirical materialist thought to why there is an evolution of the living is given himself by the fully ascertainable fact that organisms live and adapt through random mutations, without another direct purpose imposed from the outside. Today, creationist thought answers that biological evolution has a direction imprinted by a creator, God. But, according to religious assumptions of free will, this directive doesn't conflict with man's freedom. So, men could alter the course of evolution with the related responsibility towards nature surrounding them.

Regarding *how* organisms evolve, there is today a convergence of thought on the idea that evolution proceeds through random mutations of the genetic code, thus overcoming some antiquated conceptions formulated by some creationist view when fixed to simply mechanical and slavishly literal interpretations of the Scriptures.

The problem proposal

Problem proposal

An analysis is presented on the problem about how much more credible the origin of life can be when due to a fortunate and accidental combination of material events compared to the origin of life due to a generative act for God and, subordinately, if the randomness is incompatible with the idea of directionality.

The theories about how the first living being originated don't have certainties because no man witnessed the moment of his birth. For this reason, it is not possible to demonstrate how the origin of life from the elements of the inanimate material world took place. We can therefore only admit hypotheses on the emergence of life, and it is possible to analyze only the reasons for the credibility of some beliefs around this question in the knowledge that no one can, from experience, be sure of having the correct answer. On the other hand, no one can obtain special possession of the truth in many significant problems. In this regard, one can only ask for getting a logical validity regarding given premises. Questions the reason is condemned to ask itself but which, however, it cannot solve. So today, science does not know all

11

the forces present in nature but requires respect for those it does know well.

In his book, Chance and Necessity, Jacques Monod believes that the birth of life is understandable only by postulating initial conditions of pure chance and the absence of evolution future planning. He writes: "*Man finally knows that to be alone in the indifferent immensity of the Universe, from which it emerged by chance. His duty, like his destiny, is not written anywhere*" [1]. This belief, although expressed with determination, can be affirmed with passion but not scientifically proven. The descriptions offered in his book on physic-chemical and biological phenomena are correct but do not prove how life was born. This point of view, if exclusive, would seem to make it impossible to compare with other hypotheses which, with such rigid premises, would appear vitiated by recourse to ideas of pure superstition and animism.

On the other hand, it is impossible to investigate life origin starting from a living being. That is, from the already-completed living creature and subsequently studying its evolution. Since the investigation should begin ascertaining if it may exist, something as "pure chance." Because on that cognition of randomness was based the arguments proposed by Monod.

Everyone agrees that laws are not arbitrary or changeable. Only the failure to determine every event at the physical, subatomic, atomic, or molecular level makes some events unpredictable, attributing them to chance. Randomness thus removes any idea of project or intervention from the discernment of causes, as that idea would be extraneous to an uncreated and self-referential world. Indeed, one cannot even imagine a creator God who intervenes by way of miracles in every single event to override and violate the laws that he has given. Therefore, the comparison seems to be between hypotheses declared exclusively objective, linked to the idea of an absolute chance at the origin of life, and a creationist theory called subjective that starts from the idea for which a beginning cannot be otherwise than by creation. Today creationism does not exclude natural evolution, however, intended but only the birth of life from the inorganic world by pure chance alone. Of course, this means an act of believing or not believing that there is creative intelligence. For creationists, there would be a force that is also 'subjective,' which intervenes on the origin of life through a sort of metaphysics, that is, of a God who can act with other forces ontologically different, over-ordered regarding those we know sensibly or experimentally.

Regarding *why* life originated, the answer, for a current of thought that many can define as empirical and materialist, such a question is utterly improper since it would imply an intelligent entity that precedes life and has a plan for a target. For the other current of thought, called creationist, life has a meaning that by believers can be found, on a metaphysical level, in the inscrutable ends of a creator God.

About *how* the birth of life was possible, materialist empirical thought entrusts this explanation to the randomness of numerous encounters and combinations of the different elements present in the matter, leading to the emergence of the first and elementary living being. Creationist thought instead believes in a design, evidently aimed at an end, and therefore it rejects the pure chance, totally neutral and devoid of any objective and progress.

The problem proposal

For an analysis on the credibility of the hypotheses in the field, the beginning of life on earth has, fortunately, left several fossil records. So, thanks to the various finds, living forms can be attributed to specific epochs.

The primary dating method is the so-called geological stratification theory. Geologists should find the fossils of the most ancient times in the lowest strata of the earth. Today, with the method that uses the decay of carbon14, we can obtain relatively precise dating from numerous finds. However, the history of living through the fossil formations remains partly incomplete because the fossil casts that can be examined are formed mainly for living structures with hard parts such as bones or chitinous exoskeletons. Geological evolutionary testimony voids are therefore found about the first living monocellular and multicellular organisms that, mostly devoid of hard parts, such as Bacteria, Porifera, Polyps, and Mollusks and for them, easily classifiable fossil casts are not found. Even for intermediate living forms in rapid transition from one primitive form to the subsequent, due to the limited period of existence, often there is a lack of acceptable finds. However, even if incomplete, the evolutionary picture in its main lines appears today understandable or interpretable. Therefore, the biggest problem does not arise in not seeing the design of living nature in the most delicate details but occurs when trying to understand the true beginning of life, that is, for recognizing a path starting from the mineral world from the very beginning

In recent times, scientific researchers have made astounding progress in all fields of knowledge, especially in the field of molecular biology and genetics. Considering the exponential acceleration of scientific progress, the1970, the year that saw the release of Monod's book more than fifty years ago, appears a very distant period in the past. So, it is helpful to review the fundamental problem concerning the origin of life and its evolution in the light of current scientific progress again.

Therefore, it is possible today to present hypotheses more documented from a technical point of view, more advanced, and more enriched, thanks to the tremendous scientific progress achieved. The difficulties in explaining the different hypotheses arise when we try not only to describe *what* happened, like the task of classification and taxonomic analysis but, above all, when we try to illustrate *how*. That is, through what means and causes life has emerged.

Monod's text supports the thesis that the emergence of life is simply an absolute chance event because it arises from completely random initial conditions. It's about taking it or leaving it. But is this an unquestionable thesis, or is it possible that entirely initial randomness could lead to determinable and completely predictable results? In an affirmative answer, the origin of life by chance would not contradict the life origin by necessity, through a deterministic chain of interlinked causes effects to determine if life can be possible or not possible. In the latter case, life origin could be guided by external forces also in initial chaotic conditions. This direction could then occur, by hypothesis, through the intervention of phenomena that are experimentally inaccessible to man due to the hand and plans of a creator God who would act for secondary causes, through primary reasons represented by chaotic events, random but which ultimately arrive at defined and predictable results

The problem proposal

To compare opposing hypotheses regarding the birth of life would require, for a scientific mentality, the exposure and quantification of the probability of success for a first living organism to emerge by exclusively random attempts. This operation has to be compared by biologists with the birth of life out of necessity, where a probability problem does not arise but only a need to explore the means with which God chooses to operate over time. On the other hand, the mystery of life appears partly to belong to issues involving the first causes and universal principles, which cannot be addressed by scholars merely from a technical and scientific point of view. For this reason, every choice that everyone freely decides, in the face of own conclusions that no one can prove, it seems fundamental to make that decision with a reason that illuminates a belief and not with the prejudice of reasoning after a faith, blind and fundamentalist, devoid of logic.

The argumentative method

Here we talk about the method of presenting the topic of this paper and the attempt not to fall into the trap of taking for granted the truth of one opinion over another. Doubt is better than a stable and immovable prejudice.

In the twelfth century, the philosopher Peter Abelard introduced the dialectical method 'Sic et non' (yes and non; so, or not so) into the disputes concerning complex problems to solve. A confronting process led to the affirmation of a thesis compared to a denial of the same idea, using arguments utterly contrary to each other. The resulting dialectic way led to a refinement of the concept object of the debate, to the advantage of a possible truth or at least to a choice of the thesis that was closest to reality. The questions of the origin of life and its evolution belong to one of the most fascinating but challenging issues to resolve due to the absence of experimental evidence regarding the origin of life. The solution then is there, where the difficulty falls because not solvable. Eventually, not a demonstration but a choice: without

The argumentative method

prejudices and emotional involvement, one should try to glimpse the most convincing and credible arguments, relatively to the context of one's knowledge. Regarding the living world, this is to opt for a solution that accepts the thesis of an entirely accidental initial cause or for a solution that contemplates an initial and finalistic intervention by creative intelligence, immanent and transcendent

In the search for the causes that originated life, often the investigation is conducted on one lonely track alone. That is, on affirmations supporting a single thesis, as happens in many texts where is considered the existence of a pure and straightforward subjectivity either only objectivity as the cause of the nature and as the one and the only possible way. This procedure happens due to the concrete absence of a contradictory thesis. That allows authors to conclude the investigation through a mere opinion. They can state from a theoretical deduction through the creation from God or by self-creation of living nature and possible only through pure chance. In his text, Monod affirms that the absolute material randomness, present in the initial conditions, produces life and determines evolution; this idea is affirmed as the only thesis. So, he wrote: "*But once inscribed in the structure of DNA, the singular event, and as such essentially unpredictable, it will be automatically and faithfully replicated and translated, that is, at the same time multiplied and transposed into millions or billions of copies. Leaving the sphere of pure chance, it enters that of necessity, of the most inexorable determinations.*" [2]. Of course, for 'the most inexorable determinations,' he refers to the necessary reproductive fidelity of the offspring and its genetic code.

The argumentative method

The term 'necessity,' as the contrary and opposed of 'contingent,' is here wholly inappropriate as the faithful reproduction of the offspring by the parental organism is entirely contingent as mutations may occur in the genetic transmission, so parental features are not by necessity transmitted to the species. It is defined as necessary what, that without it, the essence or nature object of our attention could not exist as such. The faithful replication of a living organism is not necessary for the organism's life, and It is only helpful for the morphological persistence and expansion of the organism's species. What appears needed is that exists DNA or RNA in which to inscribe the singular event. When described as self-creative, the random occurrence of the first living being appears necessary only for a possible explanation of the insurgence of life but not, consequently, making any other potential causation that could give rise to life or evolution superfluous. This explanation, to be plausible, only requires the statement of agreement by the laws of nature with a series of random events and life initiators in a supposed infinite time available, which is possible but not verifiable, thus making this thesis a possible object of dialectics.

The thesis of a universe that does not have a temporal principle, since it has always existed, agrees with the idea of a beginning due to absolute chance because in that case, any event, even with very remote probability, can all time happen. The problem of whether the earth had been created or whether it has always existed represents, from a strictly empiric point of view, a pseudo-problem as it happens for the origin of life because of the observability conditions regarding what happened in the beginning, cannot be verified. In a correct scientific way and according to the dialectical method, Thomas Aquinas, aware of this, affirmed in this regard that he could not prove either of the two theses but that he, personally, believed that God had created the world. [3]

The creationist thesis affirms that even the laws of nature need an initial cause, which cannot be casual but necessary. While linking the zero point of the beginning of life to metaphysical laws operating within an over-ordered project, respect the physics laws themselves in the moment of creation. A dialectic way of debating crucial problems does not consist only of the confrontation of opposing theses. This way of debating requires a preliminary clarification of the fundamental concepts used in the views themselves: in our examination, we should describe what is meant by the term 'pure chance' and whether something like this exists, since if there is no 'pure chance' or if we do not define what is meant by "a necessary beginning", we can run the risk of conducting a perfectly logical and consequent discourse but starting from wrong premises.

Monod still affirms: "*only chance is at the origin of every novelty, of every creation in the biosphere. Pure chance, the only chance, absolute but blind freedom, is at the root of the prodigious edifice of evolution: today this notion central to Biology is no longer one hypothesis among the many possible or at least conceivable, but it is the only one conceivable ...* "[4] Then About creationism not being able to avoid randomness, certainly present in the primordial chaos at the origin of the earth, we need to see how it is possible to reconcile a necessary beginning of life with an initial situation of total randomness. An investigation must therefore be launched on what is meant by the term 'chance' in general, 'by pure chance' in particular, and also "by necessary or deterministic initial randomness."

THE CHANCE

M.G cm. 140 x200

You believe that God plays dice with the world; I believe instead that everything obeys a law, in a world of objective realities that I try to grasp in a furiously speculative way. Albert Einstein: Quote reported September 1944, reply to Niels Bohr.

In English, the 'chance' as a term means an anyway possibility, and it is translated into Italian as 'caso' (chance), which, at least in the everyday spoken, indicates numerous circumstances and reflects a semantic ambiguity due to two different meanings in its use deriving from two separate etymological roots. The first meaning derives from the Latin 'casus' with the primary intention of fall, what happens, accident, occasion, event, and circumstance. Under this sense, there will be expressions that concern an event. Since everything happens, this term refers to numerous expressions translated such as judicial case, police case, in case, occasion, etc. These expressions only secondarily and incidentally have a connection with random events. The latter differently should be expressed with the term 'chance' with a meaning arising from the Greek 'Týche,' which concerns unpredictable phenomena, not voluntary or finalistic, connected with the idea of a certain probability for their occurrence regarding the occurrence of all other possible events.

Most scientific thinkers agree in defining an event as random due to unknown or chaotic causes, therefore connected with the meaning root of 'Týche.' In this sense, what happens can belong to the meaning of chance in three ways: referring only to the final result defined as random because it is obtained by way of unwanted coincidences; secondly, referring to initial causation which is instead voluntary, as occurs when someone causes a possible event such as the drawing of lottery numbers, the throwing of a die or throwing the bouquet behind the bride; in a third way less comprehensible from the intuitive standpoint, it occurs when an initial cause in itself chaotic and unknowable, nevertheless causes a non-chaotic and predictable result.

In the previous work by Jaques Monod, the concept of chance is used according to the French term 'hazard,' which refers to the Greek etymological root with also a shift in meaning towards something that implies unpredictable risk: an interpretation that relates to pure coincidence as something that cannot further be analyzed. However, it seems clear that, in this sense, random events could be analyzed under their probabilistic aspect.

In the cited work, the term "necessity" clearly identifies, in a first sense, the reproductive fidelity of living organisms and does not require any probability concept to be proposed. However, departing from the sense suggested in the book, the term can refer to an original cause of life due to one creative intelligence that necessarily acts directly and voluntarily in the beginning and evolution of life through a strict rigidity of cause-effect relationship.

An event is often defined as due to chance because it is inexplicable and unpredictable; however, this definition, at least for many thinkers, does not seem to represent the truth correctly. In reality, a phenomenon would be unpredictable only due to a lack of technical possibilities suitable for knowing the previous conditions that generated a particular phenomenon. An event should then be considered unpredictable or inexplicable only due to initial conditions represented by an absolute genuine chance, and therefore not analyzable. In this case, it is entirely impossible to go back to understand the correct causal links.

The chaos theory exemplifies a situation of high randomness grade. The so-called chaos of which the theory speaks takes this definition due to minimal starting conditions that result in a process that brings paradoxical final results, thought to be chaotic, which do not appear to be connected to the initial conditions because those seem entirely irrelevant in confront of the phenomenon produced. The highly amplified sensitivity in respect to the initial conditions is commonly known as the 'butterfly effect', an effect so-called because of the title of a paper presented by Edward Lorenz in 1972 to the American Association for the Advancement of Science in Washington, DC, entitled *'La predictability: Does the flapping of a butterfly's wings in Brazil cause a tornado in Texas?'* In that paper, Lorenz describes that the movement of a butterfly's wings represents a small change in the system's initial conditions but causes a chain of events that lead to phenomena of ever-larger scale. If the butterfly hadn't flapped its wings, the system's trajectory (the sequence of triggered processes) would have been extremely diverse. In this context, however, the 'chaos' is not accidental but only represents a lack of technical possibility of final state forecasting about a system, starting from defined initial conditions. The numerically too high set of factors and variables, capable of influencing the system in its cause-effect path, makes the final state unpredictable and sometimes paradoxical in the snapshot of the final phenomena observed. In fact, in mathematics and physics, chaos theory is the study, through models of mathematical physics, of dynamic systems that exhibit an exponential sensitivity to the initial conditions but which are, in any case, not random. The sensible dependence on initial conditions means that, in a system that is defined as chaotic, tiny variations of the initial conditions correspond to highly significant variations of future behavior. In other words, each configuration of a chaotic system, under slightly different initial conditions, is arbitrarily close to another configuration with an entirely different future 'trajectory'. A real casual situation would then only occur when the initial conditions are in a state of genuine absolute and unknowable chance. Only in such conditions can it be possible to produce an unpredictable process that comes to a phenomenon in real pure random as a final result, when meant in the sense of the correct terms.

Chance, probability, and cause are also closely connected in the anthropological, social field. Trying to eliminate 'chance' as an inexplicable cause for some events has appeared to be a fundamental aspect of the primitive mentality for some scholars of anthropology.

The Chance

The anthropologist, Evans Pritchard in 1926, conducted a study in Sudan on the relatively primitive Nilotic population of the Azande. He sought to understand the presuppositions and effects of their concept of 'Mangu,' which can be translated as witchcraft. Pritchard discovered the existence of two different principles of causality present among indigenous people: on the one hand, the one aimed at understanding why a particular event happened and linked to rational explanations, on the other hand, the one that wonders why an event occurred precisely to a specific person and not to another and for which 'magical' explanations were used by them. [5]

In Western thought, we avoid using explanations that resort to magical and unscientific causes for facts that appear otherwise inexplicable. Instead, we resort to the concept of chance, which becomes necessary to explain what happens without a precise and identifiable cause. The necessity of chance is so invoked when something unexpected happens, such as happen for the classic tile that falls from the roof and hits a man: rationally we can see the root cause perhaps for the installation of the roof carried out in the wrong way or because of the forceful gust wind, but we do not find the explanation on why the tile fell on our uncle and not on another person.

Modern thought, therefore, abandons the hypotheses of magical influences regarding the causes of unexpected events but is nevertheless entangled in a problem that it cannot avoid facing but which, at times, cannot solve. On the other hand, man cannot accept such a dramatic void in our knowledge system continuity, and, as is well known, man is naturally led to knowledge. The idea of "chance," at least for some phenomena, appears to be completely necessary. This idea is considered and accepted in the sense of primitive and reasonable cause, as happens in attempts to explain the origin of life. This proceeding occurs when a scientific explanation of the causes of otherwise inexplicable phenomena is lacking. For many thinkers, however, using a so-called random cause as a reason is acceptable only provisionally, pending that science can explain every possible origin of phenomena in the future through a logical demonstrative process starting from defined causes.

The difficulty in finding definite or reasonable causes to explain the beginning of life on earth arises because of the distant time of that beginning. It has been estimated that life had an origin in a particular remote period, estimated at four billion years ago. Indeed, an evolutionary path proceeds from the most uncomplicated living beings to the increasingly complex and competitive ones. Therefore, the difficult point to understand does not consist of understanding 'what' undoubtedly happened, that is, the emergence of the first living being, but 'how' and through what causes. The problem consists of understanding which methods and means one aggregate can become a structure passing from the mineral world to the living world. That since each passage of the simple material state to another much more complex certainly implies an inevitable construction process, which must be established as the first helpful building blocks. A plausible explanation would both theses serve: either in the hypothesis that this passage is accomplished by necessity or that this happens out of pure fatality. For the event of the beginning of life, as for any other event, the way to obtain this result is due to the existence of a heuristic when understood in a broad sense. The concept of heuristics is generally connected to the search for a result and, in this case, the emerging of a first living being from the inorganic world. The term 'heuristic' is derived from the Greek language, with the meaning of 'to search' and 'to find.' This term could indicate the beginning of a research process that can be defined by necessity and deterministic considering a desired result by an intelligent being but not necessarily. So, an impersonal search could be outlined by "chance," which, like any physical process, meticulous thinkers can express in the form of probability in obtaining an unwanted result. This point of view presents two different concepts of "cause," showing two other methods for getting results through a process that can be elementary or very complicated, as it is in some scientific research. The example that is often invoked to illustrate the chance of a primal heuristic as a search method for obtaining a result, concerns about a mouse trapped inside a maze which has many ways out and of which only one will lead to the cheese room (an example: cheese is not the most popular food for mice), whose smell permeates the labyrinth evenly. In these conditions, if the mouse does not have a memory, it will implement the most straightforward heuristics to find a way out, that is:" by trial and error, without storing the negative result in memory." In this case, the little mouse doesn't remember the mistakes made, he may find himself facing the wrong way out several times, and his chances of success will be meager. If there are not many ways out, compared to the correct way one, the little mouse will probably come to eat the cheese after its blind search, trying and trying numerous times. To avoid a finalistic aspect, not contemplated for a genuine 'absolute chance,' it is necessary that there is not even the will of the mouse to leave the labyrinth since, in this hypothetical experiments, it can also be imagined by a researcher that the mouse is not hungry and therefore would have no reason to do that. The mouse should possess only a motor activity that leads him to repetitive, aimless, and equiprobable paths between them to get out of the labyrinth. This type of situation, for the mouse, will dramatically lower the chances of exiting the maze. He will almost certainly starve and casually walk through any available space without necessarily orienting himself towards some exit, lacking motivation. When would have reached any door, he could go back as he had arrived, getting lost in an absurd coming and going.

A heuristic appears necessary to explain the achievement of a particular situation or result, even when this is reached unintentionally despite the etymology of the term heuristic, which could suggest an actor who wants to achieve a result. Of course, a heuristic that must lead to a specific outcome, which in our chance concerns the origin of life, have to be profoundly different in the hypothesis of creation by an intelligent entity compared to the theory of life origin due to absolute chance.

The necessity of the chance

Lacking a demonstration on the origin of life, for materialist thought, it's necessary to find an explanation that cannot imply a divine cause. It is so necessary to introduce the idea of a particular kind of cause called 'absolute chance'.

The essential condition for imagining the birth of life through pure chance is that there may be an absence of any heuristic. Even the most elementary: simply a non-heuristic process, that is, the absence of the possibility of a research method. The creationist hypothesis loses much of its strength if it is shown that there is the possibility of existence for something that can truly be defined as "pure chance" and, as such, cannot be managed by any creative intelligence for any project or prediction. This definition of pure chance does not intend to specify the 'chance' understood scientifically, that is, as an absence of information or technical forecasting skills. Therefore, the hypothesis of the casual birth of life needs to place the "absolute chance" at the foundation of its propositions.

The initial critical condition for a dialectical process is that at least two theses exist to be compared to each other; otherwise, the only one that remains is assumed to be proven. Another fundamental requirement for the theses that are compared is that the propositions that compose them are not ambiguous and are possibly simple. For a hypothesis concerning the origin of life through God's creation, the generating cause appears clear and does not seem to be further analyzed: God is conceived as the first cause, omnipotent and person. For the opposite hypothesis, that concerning the origin of life by pure chance, it will be necessary to clarify what the term 'chance' means in that hypothesis: it is a randomness that can be seen as absolute and therefore not subject to laws and genuinely unpredictable? Or simply relative because it is subjected to the laws that govern all nature, thus a cause that allows us to predict the outcomes produced?

Jaques Monod, in his work, means chance as absolute, that is, as the first unpredictable cause of the onset of life. In reality, in his book, an analysis of the concept of chance is not carried out. Still, the idea of pure chance is taken for granted, with this simply affirming that life arose by chance by relying on a descriptive sense of the term, considered entirely intuitive and axiomatic. The probabilistic nature of the concept of the chance taken in the sense of Týche or of Hazard emerges in a few lines in his work when the low probabilities of the origin of life are affirmed as due by chance, so when he describes the fortuitous strangeness of some encounter as it can happen to everyone.

In a strictly cognitive sense, about games of chance, everyone is interested in quantifying the chances of success for a particular event or in knowing the probabilities for releasing a specific number at the throwing of the dice or for one number extraction in a bingo game. No one is usually delighted with the generic statement that success is unlikely or very likely. Therefore, the decision to take any action must rest on information relating to the quantification of success for the wished events. Regarding the problem of the beginning of life, different from the theory of evolution, a descriptive note on '*what*' could have happened does not seem sufficient. The most intriguing concerns regarding '*how*' such an event could occur and, above all, with what probability of happening.

The necessity of the chance

In the presence of two hypotheses on the origin of life, equally indemonstrable, two different deities must possibly be put as the first cause: one that posits the pure chance as a creator or another that sets the God of religion as the first cause, also the creator for definition. The idea of God as an explanation of the beginnings of life needs only faith, even if placed with the protection of reason as a "preambula fidei "(a preliminary faith), to be able to believe in a mysterious miracle of goodness. The idea, instead of a beginning due to absolute chance, if that exists, it needs to show in support for its credibility a quantification of the probability of success for the particular event of life origin. However, if the chances of success for that are almost nil, then an equally symmetrical to creationism, 'miracle' would be required and therefore the recourse to another faith.

Euripides reveals this antinomy in a fragment of a lost tragedy, the Ipsipile:

"Either mortal thoughts or vain to err of men.

As they make to be at the same time

and the Týche and the gods.

Because if there is the Týche, what need is there for the gods?

And if the power belongs to the gods, the Týche is no longer anything. "[6]

The concept of 'chance,' when understood in the sense of cause, when it comes to describing and explaining some physical and chemical phenomena that are fully defined and measurable in mathematical terms, appears necessary. This idea of cause happens, for example, in

31

the determination of the pressure of the gases, which is entirely measurable even if the pressure is caused by a total and random dispersion in some space of gas molecules. It happens because the measured pressure force is due to the accidental impact of billions of molecules on the surface where it is measured: So, only the random dispersion and equiprobability in collisions ensure the correctness and determinability of the measurement. Even the temperature, which we find homogeneous in a not large mass of water, is due to the random distribution in the movement and dispersion of the water molecules: without a random and equiprobable distribution, we even could see a part of the water mass, as in a lake, partly frozen and in another part perhaps boiling. Those phenomena, if they happened, should then be attributable to not casual and unequal dispersion and confinement of the water molecules with different amounts of movement in two lake places. And this never happens, thanks to the randomness and equiprobability of the disposition of the molecular particles in the available space when no external divisive ordering forces are applied to It. So, it appears evident that 'chance,' despite being the actual initial conditions of an event, when happening in systems with very high numbers of events or particles, can result in phenomena determined necessarily. So, entirely predictable and empirically certifiable.

In chemical reactions, we have similar situations whereby measurable reaction results can be predicted and defined in conditions that present the chaotic dispersion of billions of molecules of the reacting species. In a reaction chemical system, starting from the reacting elements, we assist the formation of other chemical compounds (represented in the right part of the reaction), and a measurable equilibrium can be seen for the chemical species present. Because of that, for an actual reaction speed, one part of the reacting chemical molecules or elements go to the right side of the reaction, and part of the chemical species from the right side goes to the left: the 'chance' appears here because we do not know which molecules, taken individually, then compose the movements that bring the reaction to a precise equilibrium. However, this unpredictability in space and time of every molecule does not imply changes in the description of the response, which appears quantitatively describable through particular mathematical equations called stoichiometric.

The elements of a system, if free to move, will always arrange themselves in the arrangement that contains the least possible energy, and this arrangement is the random one. A measure of the random distributions of elements is an implicit aspect of the second law of thermodynamics called 'entropy' which is also expressed as a measure of the disorder of all closed physical systems. Inside those kinds of systems, the disorder condition always tends to increase if there are no energy contributions from outside the systems. The Kelvin-Planck statement states that: *"No heat engine that works with a cycle can absorb thermal energy from a tank and produce an equal amount of heat."* Therefore, there is always a loss of work quantity that a system could carry out in each work cycle developed in a closed system. This phenomenon happens of necessity because the randomness must lower the system's potential energy in the arrangement of its elements, which increases the disorder. If the universe is considered a closed system, its death is inevitable. Since when the maximum possible chaos has arrived, or in other words, there will be a completely random distribution of the elements, then there will be no more energy possible.

The necessity of the chance

It is noted that the presence of potential energy presupposes some difference between two different physical states.

In a closed system with dispersed and completely random elements, it is necessary to bring energy from the outside to obtain something that is not random but somewhat structured. For a mother, tidying up a messy child's room is an obvious consideration, and unfortunately, the space will mess up repeatedly, increasing the local level of entropy. Therefore, it is possible to foresee that in each system, the spontaneous randomness in the arrangement of the elements constituting the system itself always tends to increase.

Anaximander and other philosophers have imagined an inverse, anti-entropic path, whereby the cosmos would emerge from primitive original chaos toward a structured and ordered universe. This view is now subject to the objection that it is possible to produce an ordered state of internal elements only after external energy is input into a chaotic system. But, the change of state towards a more significant order can only be conceived if, beyond energy supply, the passage will be guided in a programmed and rational way by an intelligent being: at least a creator of the laws. Chaos creationist idea supporters could conceive a hypothesis that imagines a solution for a random and probabilistic construction of aggregates that evolve into increasingly complex structures in trough times. This solution works, putting the energy problem in brackets, considering it irrelevant in local events when supposed to be full of energy. So, the formation of a structured object may be accomplished, deeming that a simple mathematical issue of probability. It is about to consider the random construction of a certain number of ordered dispositions from aggregates that interrupt

The necessity of the chance

the arbitrary distribution of elements inside the universe system. Similarly and amplified, the problem arises by passing from a set composed of innumerable molecules, scattered randomly, towards forming an organized and self-replicating structure like a living organism.

Finally, the hypothesis that presents an ordering intelligence in a world that has always existed, as for Anaxagoras (In Aristotle, Metaphysics: criticism of naturalist philosophers), in respect to the hypothesis which represents the same but in a created world and the alternative theory of the 'pure chance' as the creator of one's order, does not, therefore, leave any man no choice but to look at the reasons for credibility among the various hypothesis.

Absolute chance and objectivity of nature.

Absolute chance and objectivity of nature.

'How dare we talk about the laws of the chance? Is not chance the antithesis of the whole law?'(Joseph Bertrand, Calcul des probabilités, 1889 from The Monist, Vol. 22, Number 1, January 1, 1912, page 31.)

Where it is shown that the absolute chance, a presupposition of the objectivity of nature, cannot be conceived in the presence of laws but that without laws the principle of identity does not exist and, consequently, pure absolute chance is pure nothing.

Jack Monod, in coherence with his conception of the origin of life by pure chance, as a hypothesis necessary to support true objectivity of nature, states: "*Pure chance, chance alone, absolute but blind freedom, is at the root of the prodigious building of evolution*." [7]

However, the existence of an 'objectivity of nature' implies that when referring to the origin of life and the evolution of living organisms, events must occur purely by chance and without purposes necessarily subjective. Nevertheless, the theory about the occurrence of a 'pure chance' as the cause of the first living being can only be conceived in hypotheses regarding events not subject to natural laws, which otherwise would make them predictable and therefore achieved not by truly random but determined through rules. On the other opposite point of view, laws are necessary to support the idea of a purpose to obtain something; so, there should be a legislative and operational possibility for a subject to initiate and manage the events to be achieved. Otherwise, it would be like someone who wanted to drive a

Absolute chance and objectivity of nature.

car in which the law of cause and effect does not operate: he tries to start the engine, and maybe it starts, but it is not known if it will continue to run; it engages the gear, and the car does not move or starts, as well as for the steering wheel that perhaps turns the wheel or maybe not, and so on.

To definitively eliminate from the horizon of speculative thought any hypothesis concerning a subjectivity in nature, that is, the idea of a God who plans and directs his creation, it is required that there are no stable physical laws. Otherwise, it will always be possible to establish a cause-effect relationship between the different events, producing a necessary, determinable connection and, consequently, some possibility of direction and manipulation regarding the origin and evolution of living organisms. The presence of laws that govern nature is the object of the fifth demonstration of the existence of God presented by Thomas Aquinas, who writes: *"The fifth way is inferred from the government of things. We see that some things, as physical bodies, operate for an end, as it appears from the fact that they always or almost always work in the same way to achieve perfection: whence it arises that not by chance, but by a predisposition they reach their end. Now, what is devoid of intelligence does not tend towards the end except because it is directed by a cognitive and intelligent being, like the archer's arrow. There is, therefore, some intelligent being from whom all things have been ordered to an end: and we call this being God "*[8].

The causal determination due to the physical laws operating in nature renders the idea of a "pure" and primary chance at the basis of the origin of living organisms inconceivable. The pure chance as a causative event in this probably refers to some initial conditions, which

Absolute chance and objectivity of nature.

we do not know or cannot know while obeying these to the physical laws present in our universe. For this idea, the fact that we cannot know the initial and random conditions in the arising of life would therefore only concern an intrinsic limitation of human nature or the scientific progress of the moment. In this perspective, therefore, we would not have to deal with 'pure chance,' which, in the premises described, would be in logical conflict with the idea of the empirical existence of causal links, universally ruling nature. The logical conflict would disappear only in demonstrating the existence of something that can be defined as 'pure chance' and coexisting with the laws of nature.

In the abstract, regarding an investigation into the possibility of an actual absolute chaos situation, one can imagine a model in which what we could define as 'pure chance' dominates within a closed system. This system, by similarity, could be represented as a particular mixture of inert gas. Individual particles have simple physical attributes but without specific forces of attraction or repulsion and interact through non-preferential collisions towards other particles. In this situation, an equiprobable encounter between the various particles present in the system would arise. That would not allow us to predict anything about the circumstances produced. Since the particles in the system have no physical reason to be on one side or another in the space allowed inside the system. In such a model of pure and absolute chance, to find a critical way that over time leads from chaos to the aggregate matter and from the aggregate to a functional structure, therefore for a path consistent with the possible creation of 'something,' it would be necessary that after a stage of pure chaos, specific physicochemical laws of aggregation can intervene.

Absolute chance and objectivity of nature.

In a hypothetical universe, composed of particles with a mere and intrinsic duality of particle-motion state but devoid of other laws, one could not witness any phenomenon. Without rules or forces of aggregation, an object with a proper identity could not be born. In such a hypothetical universe, a realm of absolute chaos, the value of entropy would be infinite due to the total disorder caused to the natural and random dispersion. The complete randomness would also lead to the lack of any potential energy that anyone can use to form something. Because this would be possible only starting from any potential energy state differences between the particles or sets of particles that constitute this system. In this hypothesis, only in a sort of 'indifferent matter' could logically exist and consist with an objective beginning of the life as stated by some theories, that is in a truly and completely unexpected way.

Consequently, to generate an ordered autonomous existence of something, it would then require a different and external force operating on the system. That is an ontologically diverse force intervention that can give aggregation and order to this indefinable and undetectable matter. The example we have in our definable experience comes from closed systems, where are conditions of random distribution of molecules in gases and liquids. In those systems, we cannot witness, lacking external forces, any visible aggregation such as blocks of gas or liquids different by the mass of surrounding molecules.

For an observable phenomenon to exist, it is necessary to consider another aspect resulting from a primitive situation of absolute chance, which concerns the nature of time. Anyone who wants to measure time can realize that to detect and estimate or measure the

Absolute chance and objectivity of nature.

elapsed time, in a somewhat manner, there must be events: the movement of a clock pointer, the flow of sand in an hourglass, the vibrating of an atom, etc. Without defined occurrences, the detectable and defined duration of the period between one event and another is eliminated. A long and perceptible term between one state and another of this random matter is impossible for total randomness definition. It presupposes the existence of a non-equiprobable, non-random duration. In some way, one too long or too short period out of success probability. The consequent lack of events for which the definite time exists also leads to the fall of the principle of identity at the basis of the recognition of everything. This noetic, non-discursive principle, even if not explicitly shown by Aristotle, is latent in all its logic and is linked to a time of permanence of any object that must be present to us in one necessarily definite time for to exist as a phenomenon.

About the absolute randomness that we have analyzed, we can see that the constituent parts of a set with these characteristics do not have a definable permanence; how are those instead linked to a stable condition. So, relatively to the equiprobable continuous state change among all possible physical appears impossible to fix an observable subsistence situation. So, no significant phenomenon can be detected from the surrounding context, as it also happens for the random distribution of gases and liquids, already taken as an example. Thus, it happens that the hypotheses of a pure chance, to support absolute objectivity of nature, on the contrary, can paradoxically lead to consider as necessary the intervention of a creator, God, who introduces available energy and the laws of aggregation for the primordial chaos.

Absolute chance and objectivity of nature.

When we give up an explanation of the origin of life through a causal process of a creationist type, we must hypothesize a world with a beginning 'causa sui', a self-creation as described in the philosophy of Baruch Spinoza, a universe existing from a moment unknown to us. A scholar should take this idea without any explanation that would be impossible to formulate regarding an axiomatic principle before and outside any logic. Also accepting, for a moment, this hypothesis there would be some initial conditions, theoretically knowable, as in all physical systems. The valid 'zero' point could be seen by going back along the causal chain of events that have occurred since, in any defined causal process, a specific set of final conditions always incorporates the aforementioned initial conditions. Of course, the shorter the historical path of the system under consideration, the greater the approximation in the prediction of the final result. So, even in a world created "causa sui," the origin of life remains theoretically, however, predictable through the analysis of the deterministic processes in progress dictated by the unavoidable natural laws. Even in these hypotheses, it is entirely absurd to speak of pure chance, when this is understood pure in the strict sense and, also, in this case, we must instead consider the term 'chance' only in the broad sense, a term used for the lack of technical knowledge of the material elements and of the forces at work in the systems analyzed. When we later speak of an absolute chance for some situations or events, this term will be understood broadly as unpredictable results because the system's initial conditions to be analyzed are simply unknown.

Donald T. Campbell, in his presentation of a theory of knowledge, modeled on the evolution of living forms called "evolutionary epistemology," was aware of the non-applicability of the

Absolute chance and objectivity of nature.

concept of chance when understood as absolute. So, he preferred to use the terms 'blind variation 'instead of the term 'chance.' That considering the absence of purpose, as we know it in humans and higher organisms, but not the absence of laws about other events otherwise called random. Many biologists then used this terminology to illustrate an analogy with natural evolution in the growth of knowledge. It was an analogy based on the mechanisms present in the biological evolution of the acquisition of new forms through random mutations. The evolution of knowledge for this theory is seen to operate through an unintentional heuristic process for 'trial and error with the memory of the error,' considering the proof, hypothesis or theory, almost 'random' and the error then discarded from the history of knowledge thus obtaining selective conservation of the result, that is of the theory best suited to survive, so as that happens for living organisms in the biological field. [9]

Then for reasons of descriptive economics, we speak of randomness, especially for biological phenomena, meaning randomness in this sense, thus distinguishing between a situation of "pure chance" as it is quite different from a case of chance or chaos instead ruled by laws.

The deterministic randomness

The deterministic randomness

Where it is shown that a random initial situation can lead to a non-random but deterministic result because it turns to be necessitating by the nature of the universal physics laws.

The final result of any physical process can be known and governed if the given initial conditions are known. However, in a situation of pure chance in the initial conditions of a process, it is assumed that it is impossible to predict anything about the final approach to which one results after some time. Therefore, not even a superior intelligence would have a practical possibility of intervening to determine the result for any desired project. Consequently, absolute chance in the starting conditions of any physical process, eliminating the possibility of addressing the desired effect, would exclude the validity of the idea of a creator God. That, except for the hypothesis of an ordering and legislative intervention of God, that, however, should be accomplished after the pure initial chaos. But this is not what those who declare life to be born of pure chance intend. The theory of the origin of life through pure chance alone, in fact, implies the absence of a creative intelligence entity. In contrast, the intervention of an intelligence entity in the creation of life would exclude, in any case, the chance, as represented in Euripides (see page 19).

"... Because if there is

the Týche, what need is there for the gods?

And if the power belongs to the gods, the Týche is no longer anything." [6]

45

The deterministic randomness

But is this mutual exclusion proper? For Monod, yes, in his book, he affirms the origin of life through absolute and objective randomness, excluding in his opinion any hypothesis of a creator God. With this, he asserts that chance can only generate unpredictable and therefore also indeterminable effects, confirming the fragment of Euripides in the part of his question that excludes the intervention of God or gods. However, upon careful analysis, this mutual exclusion appears unfounded because it is based on the possibility of situations of absolute chance, which, as already seen, is not physically possible. The problem then remains in establishing whether a position of pure chance, in the presence of physical laws as commonly accepted, can allow the generation of life, but not excluding the possibility of the work of a creative intelligence entity theoretically. If it is assumed that even completely random initial conditions in a physical process can be deterministic causes for final events, then correlatively, this idea it would show that life can also originate from initial conditions of "total chance" without thereby eliminating the notion of intervention of a creative intelligence entity. So, supposing chance to be a cause that determines a specific outcome then would be possible by God to carry out a project that is not necessarily linked to an initial order that is entirely pre-established but casual, on the common sense of the term.

Everyone agrees that if we were able to follow the fate of every single particle, even the most elementary, of our universe, we would be able to eliminate the idea of randomness in the occurrence of events. Indeed, knowing every possible cause-effect relationship would allow us to predict any event correctly. From a strictly scientific point of view, it is natural that every observation must be carried out through specified investigative operations, including the observer's role and position, since man does not have an absolute point of view valid for every situation from which to judge and relate every particular event. The current operational theory of science does not allow for every scientific activity to escape from the need to describe the mental, experimental and physical operations connected to every possible experience along with the own observation point. Describing the fate of every event, even of a tiny nature, does not escape this necessity even if, in practice, for events at the subatomic level, a complete description in this sense does not seem to be possible. The Heisenberg uncertainty principle excludes the possibility of simultaneously determining the dimensions of time and space for particles of minimal sizes. The observation in those cases would be carried out with probes containing energies capable of perturbing the observed object, just as it would happen if we wanted to draw the contours of a hair using a pencil.

On the other hand, even if it were possible to follow the path of each particle, since the number of atomic and subatomic particles and the interactions between them is enormously high, it is not possible, at least today, to mathematically calculate the fate and effects of a single particle on numerous particles. In determining events in atomic physics, it is not stated that a particular single particle does not have a defining moment of time and space but only that it is impossible to measure it or, precisely, to determine it.

Leaving the specific field that refers to the orders of measurement of single atomic and nuclear particles, we can nevertheless see the randomness, in its effects, as a measurable variable. Erwin Schrödinger observes: "*In general, a variable does not have a defined value before measuring it; therefore, measuring it does not mean ascertaining the value it has. But what does it mean then? There still has to be some criterion for whether a measurement is true or false, a method is good or bad, accurate or inaccurate - whether it truly deserves the name of the measurement process ... Now it's pretty clear;* <u>*if reality does not determine the measured value, at least the measured value must determine reality*</u> *- it must actually be present after measurement in that sense that it will be recognized again in an independent form. That is, the desired criterion can be simply this: the repetition of the measurement must give the same result. With many repetitions, I can prove the accuracy of the procedure and show that I am not just playing. It is good that this program corresponds exactly to the experimenter's method, to which the 'real value' is also not known in advance.* "[10] (T.d.A. underlines not present in the original text

Even the most elementary particle does not possess absolute freedom, as an event due to actual and complete randomness should require: since everything is still subject to the laws of physics that govern the universe. The rules, in turn, escape the definition of randomness since they are utterly immutable, as evidenced by the regularity in nature. These laws are therefore also the cause of the 'random' distribution of particles that certainly do not escape the constraint of physical laws, which then allow to describe and mathematically predict the outcome of phenomena due to events even when defined as random at their origin. However, this occurs only for phenomena where numerous elements or events are involved, therefore following large numbers' probabilistic law. In this situation, the randomness in the distribution in the space and the events produced by single particles becomes necessitating, making the final results determinable and predictable of their physical state. The phenomena so will leave the realm of contingency, that is, of what may or not maybe, to enter the kingdom of what must necessarily be in a certain way. As stated, this occurs for phenomena with orders of magnitude much more extensive than those affecting small numbers of single particles, and this is a widespread event in nature. It is known, for example, that a single gram molecule contains, according to the Avogadro number, many molecules that amount to $6.022 \cdot 10^{23}$, that is, trillions of billions of particles in a few grams of matter

The determinability of physical experiences originating from vast numbers of particles, with an order judged to be random or disordered, is evident above all in the phenomena that directly involve the second law of thermodynamics, which describes a spontaneous increase in entropy in any system when not subjected to external energetic influences to increase inside the order. This law applies to the description of the pressure of a gas, the temperature, and chemical reactions. However, in those cases, it is not essential to follow the history of each atomic particle but are relevant only to the physical laws regulating the movements of the individual particles. According to the law of entropy, then there are entirely random movements and distributions of particles which, as a whole, can be defined as necessitating, deterministic randomness for the phenomena produced; that is, they have predictable phenomena, with a non-random but a determined outcome.

Events arising from deterministic randomness can be subject to mathematical treatment, unlike events due to acts of free human individual choice. In numerical terms, the number of occurrences taken into consideration distinguishes two different ways of operating for the

The deterministic randomness

'chance' when seen as a source of variation and origin for the phenomena that occur in the world. The number of objects present in the initial conditions of the events, which will give rise to a particular phenomenon and the frequency of the events themselves, decides whether the 'chance' can result in unpredictable phenomena or predictable phenomena. This last form of chance, as already described, is "deteministic," therefore capable of hesitating in deterministic and necessary terms precisely because it is random: in such cases, each variation of the single-particle will find billions of other particles with equal and opposite values, ultimately resulting in a fully defined mathematical average.

In the evolutionary process of the living, even the mutations that allow evolutionary leaps from one species to another may appear wholly random but, somehow, definable. An example is presented in nature by the process described as 'biological convergence': a phenomenon in which the evolutionary processes of different structures are channeled towards similar shapes and forms through entirely random mutations for each evolutionary line. Many modifications have a somewhat defined path, as they are conducted and inscribed in the possibilities of the genetic code. That places restrictions on the infinite possibilities of a purely mathematical combinatorial such as that which is subjected only to the laws of physics and chemistry without considering lethal mutations and therefore impossible to transmit. In the probabilistic event of the origin and development of life, the modifications that occur in large numbers rarely produce healthy living beings; since they often came out of the legislative possibilities of living nature.

The deterministic randomness

In some way, especially in genetic fields, when again dealing with large numbers, the randomness consents the possibility of making predictions on a gene pool when it comes to events for the distribution of genes in vast populations. Hardy and Weinberg's law allows us to calculate how the evolution of gene frequencies appears in a population of organisms. This law can be stated as follows: *"In a panmictic population at equilibrium, (population in which crossings occur randomly) where neither selection nor mutation occurs and is composed of numerous individuals, the ratio between genes (alleles) and genotypes is constant from one generation to the next."* [11]

Another interesting example concerns the phenomenon of the rate of gene mutations per generation, which is considered constant with a certain approximation. This rule applies to the different species and genera, in which each species has its rate of mutation over time. That is a phenomenon used in population genetics to calculate epochs when two different species differentiate from a common ancestor and form two distinct lineages. The higher the density, understood as the quantity accumulation relative to gene mutations, and farther in the past is the divergence in speciation from the shared genetic progenitor. That determines the point in which genetic differences have become so profound as not to allow members of animal populations to be interfertile even if they originate from a common ancestor. In the presence of mutations that occurred in a completely haphazard way, it is, therefore, possible to have an approximate biological clock, but also sufficiently precise if placed relating to the geological times under consideration, and thus predict the period of essential separation, of species, between two different genotypes.

The deterministic randomness

The possibility of tracing a certain number of gene mutations to the time of speciation is essential to us from a scientific point of view. For example, it has been calculated that between 1,400,000 and 900,000 years ago, Neanderthals and Denisovans, the first ancestors of modern man, were genetically separated from a super-archaic ancestor. Between 1,000,000 and 800,000 years ago, Denisovans and modern man separated. In the range of 700,000-550,000 years in the past, the Neanderthal diverged from modern man. However, there is evidence that Neanderthal and modern man could be to some extent inter-fecund and, for the delineation of species which defines different species each other non-fertile populations, the genetic separation appears to some extent progressive. Even though the Neanderthal lineage became extinct 39,000 years ago, the European genome is 2/3% Neanderthal. [12] These dating show that entirely random events, such as mutations, can be used to compute relatively exact historical reconstructions.

Considering that randomness, often, could be a subject of accurate predictability in terms of statistical mathematics, the clear contrast between chance and necessity appears when we consider randomness as a condition attributable only to a particular individual's singularity, for which we cannot predict the result of convictions or volition. It is possible when this individual singularity is endowed with a free will, as is presumed for man, and not otherwise discernible to an external eye. In other circumstances, the contrast between chance and necessity appears real only by altering the concept of what comes to be considered 'chance' which, taken as such, is simply a helpful definition to mask our incapacity and the technical impossibility of the knowledge's period state on the underlying phenomena resulting from a specific root cause.

The deterministic randomness

The mathematics of large numbers states that it is not impossible to have a form of efficient cause capable of operating in a deterministic way even through initial conditions of total chance. So, a particular state of randomness does not deny creative acts given the predictability of the effects: which can be a cause in all respects as other causes, defined non-chaotic because they can already be known and determined from the start. Therefore, a divine creative intelligence can operate and foresee results even through chance and finds its limits only in what is unpredictable, such as the human will. Of course, for an omniscient mind, nature cannot proceed except as it does, through the laws and created things at the beginning. Indeed, in this objectivity, different from nature objectivity due to an impossible pure chance, a free subjectivity can nestle because of humankind's free will, which breaks the regular flow of events established since creation or eternity in phenomena consistent with objective determinations.

The subjectivity of nature in the laws.

The subjectivity of nature in the randomness regulated by the laws

It is shown how the hypothesis of the subjectivity of nature does not exclude a determinism in the proceeding of living nature and does not exclude the possibility of a free and unpredictable will too.

The conception expressed by the philosopher Baruch Spinoza to demonstrate the beginning of the universe affirm a monolithic principle devoid of references to other causes about the universe beginning. He defines this principle as "causa sui" (that it is the cause of itself), and it is so based on initial self-determining conditions. Spinoza presents a conception that seems different from both creationist hypotheses and those linked to a world without initial causes, both for a subjective conception or objective conception of nature. The world thus defined is then assumed, in its modes and attributes, as being God himself, entirely immanent to nature because He is ultimately the totality of nature itself. This conception distorts the concept of God, who, at least in monotheistic religions, is also a person and not just pure immanentism that identifies with the things themselves. From the affirmation that nature itself is God arise the strange definition attributed to Spinoza's philosophy as a pantheistic philosophy, derived from many thinkers' opinion that everything is God. At the same time, God has thus been eliminated by these believers; since a god conceived in this way appears to religion not present except mutilated and by homonymy.

As stated, a world governed exclusively by physical laws, as we know them, excludes, theoretically, the existence of a situation that can

The subjectivity of nature in the laws.

be defined as 'pure chance' since phenomena and events remain, thanks to these laws, however predictable. Such a world can be imagined as an enormous cosmic clock in which, given the initial original conditions, everything takes place in a deterministic and mechanical way without man being able to do anything different.

Starting from a dogmatic and incorrect premise, that is, from a god who is not God, one can draw logically coherent and grammatically correct conclusions, which are nevertheless not true. Spinoza, in his original text, "Ethica more geometrico demonstrata" at the end of each page, always wrote the acronym QED: "quod erat demonstrandum" (as demonstrated.) This affirmation was entirely correct, except for the premise. Of course, lacking a real operating subject, the idea of nature in Spinoza cannot be considered as a conception of subjective nature.

Many phenomena in nature, unlike mathematical logic, cannot be proved as specific final results of a causal chain starting from an absolute principle. The occurrence of a phenomenon expected can, by us, only be expressed in terms of probability. This kind of situation occurs, for example, when we bet on dice on a number in which except for the presence of cheaters, all events are equally likely, and we expect to win once every six rolls, and this then happens in reality if everything takes place with a high number of dice rolls. The same example applies to waiting for the release of the first number drawn in Bingo, in which we have one chance of success out of ninety. Those who hypothetically possessed particular cognitive abilities such as a powerful calculation capacity, complete knowledge of the initial conditions, as well as the underlying physical processes that occur before the results would see everything that happens in a definite and predictable way. According to

56

The subjectivity of nature in the laws.

that idea, this would leave no room for hypotheses for an actual free choice in interpreting the total objectivity of nature. Instead, a rigid procession of cause-effect relationships would link everything.

The problem concerning the freedom of the human will has been much debated in the history of thought. For theological questions, too. In his dissertation "The freedom of the human will," Arthur Schopenhauer exemplifies this conception, whereby everything happens for an initial cause that he calls "principle of sufficient reason." The motivation for naming the concept of cause in this way serves to clarify that cause-effect relationships do not belong only to the world of physical events. The principle of cause is so expressed, with a more suitable term for showing events also produced by the will of man. Schopenhauer affirms that if there is a sufficient reason for it to happen for every event, the will of man is not free since there is always a reason behind his decisions. With this, he supports the idea of the servant human will against the concept of free will. He clearly expresses this idea and writes, "But to bring these hesitations to the right topic one must ask in the following way and not turn away from it: 1) To a given man in given circumstances, are two actions possible or only one? The answer of all those who think deeply: only one. 2) Considering that the character of a given man is fixed and immutable and the circumstances of which he was influenced were necessarily determined, all of them, down to the smallest, by external causes which always occur with rigorous necessity and whose chain formed only of equally necessary rings, salt to infinity, could his life be different from what it was, albeit in some event, in some scene? No! Here is the logical and correct answer.

The subjectivity of nature in the laws.

And here is the consequence of the two theses: everything that happens, from the greatest to the minor fact, necessarily happens" [13]

According to this interpretation of the concept of cause, Julius Caesar could not have doubts. Yet, he was instead bound, without the possibility of choice, to the causal chain that had led him up to that moment when he wanted to cross the Rubicon? With the famous phrase: "Alea iacta est," translated in the historical texts with, 'the die is cast,' Caesar was undoubtedly aware of deciding that entailed a not too low probability of success compared to a failure. In that chance, he had reason enough to take the risk and, from a strictly rational point of view, he had to choose to cross the river.

The argument that can undermine the conception expressed by Schopenhauer is that the significant problems concerning God, soul, and world which, as Kant states, reason cannot solve but cannot avoid facing. Then are not decidable only in a probabilistic way by reason or own particular temperament. Still, they are original choices subjected to freedom of choice for which probability, as a determining reason, is overwhelmed by completely subjective considerations. That may be based on events that have not happened or have not been experienced. This freedom of choice, if realized, would represent the "absolute proper chance" that cannot be determined in its effects, with radically unpredictable results as implicit in the concept of free choice.

Suppose, instead, that we admit that mental processes are predictable like any chemical-physical reaction, thus disputing the existence of an unpredictable free choice in men. In that case, one could not contradict the pure objectivity of nature. A pure objective determinism would reign in the world of physical events because the

The subjectivity of nature in the laws.

concept that concerns true randomness for events is connected with the idea of absolute unpredictability. Meaning that any possible observer could never foresee something starting from particular initial circumstances because characterized by conditions of pure chance or from intrinsically unknowable initial situations. However, it has been seen that a pure initial chance is impossible; since it would presuppose the absence of some fundamental laws of the universe. Still, we noticed that only truly unknowable initial conditions could contradict the total objectivity of nature. This lack of acquaintance could only happen for humankind sort of chance because of the impotence of anyone, to read inside beyond mind surface, that is to be able to read the content of the human mind. It is not possible to know the origin of mental processes involved in free choice in a man. This impossibility arises since it is clear that the user, of his cognitive processes, cannot be removed from the point of observation of his own and particular thoughts, except to substitute another brain for him. However, even then, it would not be his thoughts that we would learn, and therefore we would not be able to know his choices. Furthermore, any possible intervention by external probes to understand these choices would alter his mental processes, as in a sort of biological uncertainty principle.

The real unknowability of the mental processes of man arises from the more general consideration of the role of the observer who must be present in any physical process to be described, as is postulated by the theory of relativity, for which doesn't exist an absolute point of view in the description of anyone physical parameter. For this reason, it is always necessary to specify the time and space to which the physical phenomenon to be detected refers. An example consistently reported illustrating relativity present in the physical world is the one that

The subjectivity of nature in the laws.

describes a fisher along a river who simultaneously hears the sound of two claps of thunder, one to his left and one to his right; another distant fisherman must contest this idea of simultaneity. Far from the former of a few kilometers, this latter declares that the two claps of thunder occurred in sequence and not simultaneously. All those placed at significantly different distances from each other will tell with full conviction which thunder happened, before or after. With extreme approximation and only to illustrate the concept, to be able to have experience and therefore know the phenomenon of that simultaneity, one should place oneself in the exact place where the fisherman who lived it sits, thus chasing him away from his position, but then there would no longer be the fisherman from whom we would have liked to have the information on simultaneity.

Regarding the emergence of a subjective choice, which usually is many orders of magnitude more complicated than the evaluation of a time elapsed between different events, we must therefore admit that what happens and determines a man's will depends on his interpretation of the events. In the light of his convictions: to know his choice, we should have experience of his entire history and his evaluations and how he thinks them at the moment and "in that precise time" in which the investigation to be conducted should be placed. In other words, to determine the initial and final conditions of an individual mental process, it would be necessary to be able to occupy the place of observation of the individual in his cognitive processes while choosing among the thousands of possible alternatives of reason but, to obtain this, we should have his own and exactly its specific point of observation, which as we have seen is entirely impossible. Therefore, it is equally impossible to know the formal origin for some results that

The subjectivity of nature in the laws.

consequently turn out to be caused to absolute chance because they are theoretically unknowable.

The 'absolute chance' due to free choice in man would represent only a particular analogy of a universal subjectivity of nature which indicates in God the subject who, before any object, through ontologically different forces from those we see present in the physical world, creates the universe. In this idea, these forces should govern the world through deterministic causality, entirely subjective, before making the world and following the laws of nature that are not alien to or 'other thing' than God.

According to monotheistic religions, the One who creates and governs the world frees the will of the individual man, who otherwise would not be morally responsible for what he does. The resulting freedom of choice makes the randomness truly unpredictable by presenting the intervention of a real "pure chance," randomness not deterministic and apparent as the one conditioned and dictated according to the principle of pure objectivity of nature bound to natural laws. Pure chance enters the universe only with the emergence, in biological evolution, of the unpredictability of human will, not always determined by the existing environmental conditions. The animal world, instead, in its mental processes, is linked to primitive systems such as chemical or physical tropisms or simple and conditioned SR (Stimulus Reaction) systems. Even with an initial symbolic intelligence, it does not have a capacity for abstraction that can lead to imagining a transcendence beyond the physical world as happens in man, by example, about the cult of the dead. This cult indicates the possibility of conceiving an afterlife in a time, free from the immediate sensitive world. For these reasons, animal intelligence before the man cannot produce genuinely random processes but only highly sophisticated methods that can be traced back analytically to complex SR systems in which, for the most advanced animals, the 'self' coincides perfectly with one's own physical identity. Which indicates an object, not detached from the physical context surrounding it, determines it and makes it predictable.

On the other hand, the conception of a deterministic world, devoid of purpose and entirely linked to origins of 'pure chance,' can only lead to the affirmation that man, like every other living being, consists of a simple nutritional element among countless others, simply forming part of the overall food chain in the current biosphere: a mere transit of food. Under the vision of total objectivity of nature, the whole ideological apparatus, which supports moral and social conceptions, appears to be the direct expression of exclusive biological instincts tending to the conservation and perpetuation of the species which, like any other animal, live one natural objective ethics. The logical conclusion of this vision implies that anger, kindness, gratitude, expressions of respect and love: that is, all emotions in general and their presentation, would fall within the category of biological processes suitable for survival and would therefore only be the expression of physical acts and impulses that have not become actual behavior, as illustrated by Charles Darwin in work: "The Expression of the Emotions in Man and Animals" [14]. In this perspective, emotions would be acts not completed: anger would represent an emotional state linked to unfulfilled physical aggression; a smile would express on an expressive level an affective gesture not concluded, a severe attitude represents the desire for an unfulfilled removal, and so it would happen for all other

The subjectivity of nature in the laws.

emotions or emotional states. More crudely, Arthur Schopenhauer, in his major work, "The world as will and as representation" represents nature as a place in which the will, understood by him as the will to survive, presents itself as a means to preserve the individual and the species, even within the own species just like in Latin: "Homo homini lupus," man is a wolf also to the other man. In his idea, other representations will deceive us into believing that the choice of another human being to love and the generation of a new creature is our free and spontaneous acts where, instead, it is our very will to live which, through these acts, affirms itself, assuring us of the continuity of the species. The animal world feeds on the plant world, and animals devour other animals in the universal will of survival and subsistence that operates from the simplest to the most evolved organism.

he presence of natural laws governing events makes it possible to hypothesize objectivity of nature in which the world, as a cause of itself, can be considered an 'entelechy' as a whole. Aristotle coined the term to describe a different way of passing, for anything, from a possible event to a made event, not for an external cause to the thing but thanks, its internal causality. That would seem to overcome the idea of subjectivity outside nature because unnecessary for accomplishing something.

Beyond the difficulties in accepting the self-contradictory hypothesis of a beginning of the world as the cause of itself, the idea of a universe governed by objectivity alone encounters a difficult obstacle to overcome that arises when one wants to reconcile this vision with existence, never contradicted, of the second law of thermodynamics which provides, that lacking forces external to a closed system, the final

The subjectivity of nature in the laws.

achievement of a physical state that possesses the level of maximum entropy is reached, that is, of a state with a distribution that is the as random, uniform and disordered as possible for all its internally constituent parts. This final equilibrium would result in the energetic death of the "universe" system, taken as a closed system because it is the totality without external things. That is because of the lack of potential energy differences between its parts. In a conception of total objectivity, the world would then represent an unrepeatable experience, which arises from itself and dies for itself unless postulating a force, external to the "universe" system, ontologically different but inconceivable from the utterly objective point of view of nature.

For those who hypothesize a subjectivity of nature against a conception that foresees an exclusively deterministic world, in which nothing is truly random, a different conception of a world can be hypothesized, in which another order of forces besides the perceived physical ones is represented, that is, an order of energies which can be defined as metaphysical and for which a factual subjective unpredictability or randomness can also be conceived. In other words, a different world vision can be supposed. There is true randomness, at least for a particular class of events, alongside other phenomena and events that can instead be subject to entirely deterministic events as for absolute and separate objectivity of nature. A subjectivity of nature contemplates the presence of natural physical forces as they are sensibly present to us in one way or another but also of energies, ontologically different from those known.

The idea of forces other than those known is perhaps the most opposed hypothesis by supporters of a single possible origin of life for

The subjectivity of nature in the laws.

pure objectivity. This opposition is justified by the fact that at least until a few years ago, there were no exceptions to the physical order described by the laws of modern physics, represented above all with Einstein's famous equations on special relativity and then on general relativity, which experimental observations have always confirmed. Today, however, it is possible to produce examples of forces and phenomena that go beyond the confines of what was imaginable until a few years ago. Recently, experiments have contradicted the unambiguous statements on the nature of time and space, held up to now, demonstrating the existence of an extraordinary quantum phenomenon that, until recently, had to be defined as metaphysical, called "Entanglement" (from English: tangle, intertwining). This phenomenon demonstrates the connections between particles with meanings that go far beyond the terms hitherto considered insurmountable relating to the concept of locality, time, and space.

According to quantum mechanics, it is possible to create a set consisting of two intrinsically connected particles characterized by a specific global value of the observable forces. This phenomenon implies that the value of the parameters measured for a particle of a defined property of the set instantly influences the corresponding value of the other, maintaining the initial global measures. The 'strangeness' concerns the observation that this value remains true even if the two particles are spaced apart, theoretically without any spatial limit. These strange phenomena happen, violating the assumption of special relativity, which indicates the speed of light as not exceedable. The concept of time, based on the succession of events, and space between different objects, would imply a distance to travel for distant events. Still, since in the phenomenon of entanglement, the occurrences are

The subjectivity of nature in the laws.

instantaneous and not serial in time, in reality, a paradoxical value is presented in the elementary, intuitive fundamental equation that links the dimensions of time and space, that is: speed = space/time, $(V = S / T)$ since the distance, which we also perceive, understood as the space between two parts for this equation vanishes since $S = V \times T$ with T equal to zero, and then the distance remains zero while the speed is infinite. The new quantum physics acquisitions have disputed the reference, considered indisputable, about the impossibility of exceeding the speed of light in a vacuum. Therefore, the concept of "locality" has been surpassed, even though it is an undeniable basic principle for classical physics and is called "local realism." Physical principle until recently considered entirely inviolable.

Quantum entanglement has been experimentally demonstrated with photons, neutrinos, electrons, large multi-carbon molecules (buckyballs), and even small diamonds. [15]. Entanglement phenomena in communication, computing, and quantum radar are a very active area of research and development. Some experiments have validated this theory. In 2016 at the University of Leiden in the Netherlands, the entanglement of the angular moments of four photons was ascertained [16]. More recently, in 2017, there was a phenomenon of high-distance entanglement when Chinese researchers launched the Micius satellite, which was supposed to communicate with three ground stations: those of Delingha in the Qinghai region Nanshan in Xinjiang, and of Gaomeigu. In Yunnan. In the satellite in orbit, the operating systems then produced two entangled photons separated by 1203 kilometers from each other [17]. Some researchers also performed other experiments, reflecting entangled interactions between light and bacteria, that is, in living organisms [18]

The subjectivity of nature in the laws.

In 1935 Einstein himself, Boris Podolsky, and Nathan Rosen formulated the famous "EPR paradox" (from the initials of the three scientists), which highlighted, as completely paradoxical, the phenomenon of entanglement. The term entanglement was introduced by Erwin Schrödinger (Proceedings of the Cambridge Philosophical Society, 1935) in a review of the famous article on the EPR paradox, which theoretically revealed the phenomenon. Entanglement between particles, which counterintuitively implies the presence of distant correlations (theoretically without any limit) between their physical quantities, determines the non-local character of nuclear and subnuclear relations stated from the theory.

What a few years ago was conceived as a metaphysical concept has finally entered the physical universe with manifestations until recently considered ontologically different since the implications would be:

• The structures, connected in entanglement, can act without limits of time and space.

• Thanks to the simultaneity of the relations between the parts, there can be delocalized unitary structures with elements not necessarily contiguous in space, as we have considered.

• A sensibly present material structure, a body, even when decomposed or dissolved, can theoretically maintain a shape not necessarily bound by the usual proximal Coulomb electrostatic forces or even nuclear and subnuclear ones.

The subjectivity of nature in the laws.

• The universe is thinkable, even scientifically, as intrinsically connected in all its parts even without the means of the physical forces known so far.

In the past, anyone who had shown such ideas would have been accused by the scientific community of cultivating beliefs in the odor of superstition. With this showing, in retrospect, it can be said, a preventive ideological closure towards ideas other than their own considered today, however, undoubtedly true since they can experiment and genuinely scientific.

Consistency and truth in nature.

It is shown how the hypothesis of complete objectivity of nature can only have the value of coherence but not of truth, unlike the belief that does not exclude a subjectivity of nature which can confer the significance of truth where a coherent theory about something finds experiential confirmation of the thing.

Coherence is the fundamental criterion of relationships in an exclusively objective world vision. Given the fundamental laws and interactions due to the different electromagnetic, nuclear, and gravitational forces, all the world events are expressed by coherence. Whoever embraces this theory sees the unfolding of events in the world with extreme simplicity. Ideas can represent a univocal vision of nature under a single intrinsic order and criterion for the forces operating in the various cause/effect relationships. The consistency criterion, simply coherence, differs from the truth because it is based on conventions as, according to this criterion, the events would descend directly from their axioms and postulates considered fundamental or, in any case, placed as a foundation on which to base the explanation of subsequent events: what happens is accepted by convention and does not require further reasons beyond the initial premises. The game of cards represents a good example. Given the assumptions in which it is established which cards or combinations of cards are higher or lower than others if the rules are respected, the game system in its development is consistent with the established practice in which a specific score will always surpass others, just as a mathematical theorem is invariant in its results if its founding axioms and the rules of its logical operators are respected. This criterion of coherence is also affirmed in the first

Consistency and truth in nature.

famous incompleteness theorem by Kurt Gödel [19] in his article 'On the formally undecidable propositions of the Principia Mathematica, and related systems,' prove that within a consistent formal system (intrinsically coherent) it is not possible to affirm a truth. In fact, in the first incompleteness theorem, he states: *'In any coherent formalization of mathematics that is powerful enough to be able to axiomatize the elementary theory of natural numbers - that is, powerful enough to define the structure of natural numbers endowed with the operations of sum and product - it is possible to construct a syntactically correct proposition that can neither be proved nor refuted within the same system.'* About the second theorem:*' No system that is coherent and expressive enough to contain arithmetic can be used to prove its own consistency'*. In other words, there is a need for another external system, different from the first one, to be able to enunciate a criterion of truth: in the chance of the card game as in other conventional games, each player needs to follow and respect the rules to affirm the fact, that a specific score exceeds another, just as in the arithmetic calculation two plus two equals four if one ascertains that two pebbles added to two pebbles are really four. So, an indication that a guy gives to a tourist looking for a particular pizzeria in the country can only be considered true when the tourist, following the directions, finds the pizzeria, that is, when the statement contained in the indication is demonstrated, in fact, through experience.

Any system that starts from an axiom or sets of axioms falls under the conclusions of this theorem, as does any game that constitutes its own initial rules as constructs or assumptions established by convention. In the game of poker, these rules must be agreed upon at the beginning of the games, as sometimes some rules differ from

Consistency and truth in nature.

country to country. If you change the founding rules during the game, you question who wins and losses in the reality of the game. The actual outcome of the play confirms established rules. Truth, therefore, requires a duality of different principles and laws that must operate in other ontological systems in an agreement of mutual confirmation.

Thomas Aquinas stated that relating to the concept of truth is 'adaeguatio rei et intellectus': correspondence between reality and intellect. In other words, truth requires two different spheres of existence to be defined as truth. So, only consistency can be found within a single sphere or reality system, assuming it has it. For example, I can affirm that the moon is square by enunciating a perfectly coherent sentence within the grammatical rules in the grammatical system. Still, outside this knowledge system and without an agreement with the facts, it is pretty evident that this sentence has no value of truth. From this statement, we can affirm that the idea of perfect objectivity of nature has no truth value even if it may appear consistent with the set of physical laws present in the universe, which, taken as axioms, cannot be denied in their existence as a logical entity. The idea of subjectivity of nature can have validity, therefore possess the value of truth, only when the subjectivity agrees with objectivity. The temptation of philosophy to seek a single principle to explain reality only leads to the conception of formal systems, intrinsically coherent with the founding and axiomatic principle, which is assumed as the first cause. Thus, systems of mere subjective idealism appear in the history of thought that attempt the leap from idea to being a real being, pretending not to be simply a being of thought. The thought history has seen, opposed to pure idealism theories, merely objective systems that attempt to leap from the only material and experiential world to other concrete, metaphysical too,

Consistency and truth in nature.

which requires ontologically different law as the moral laws, that are not merely biological rules for life survival.

The affirmation of an objective conception of the living world that we can experience when represented as a closed system in itself can only be consistent. Still, to be considered a real thing, it must find confirmation with another system that must be ontologically different. The conception of nature that represents among the causes of its beginning also the idea of true and absolute randomness must introduce into reality, in addition to a rational value of coherence, also a matter of truth, which we can find in the confirmation between two different systems that could reciprocate each other. The concept of a world governed by objectivity alone is found, as in arithmetic, to possess only a value of coherence. This consistency is ensured by the ontological and axiomatic foundation of the system of natural laws, certainly not unknowable. This conception and what it affirms can come to possess a truth value only when confirmed by an ontologically different reality system, which can only be that of a metaphysical subjectivity of nature. The converse regarding the conception of a world based only on pure subjectivity can imagine an intrinsically coherent but only abstract universe, without truth value if devoid of the sensitive and experiential confirmation of objectivity of nature.

The difference between a world of perfect objectivity, compared to a world in which there is also a subjectivity, is in some way also exemplified by the episode known as the dilemma of the ass of Buridan, a denomination taken from the figure of Jean Buridan (1300-1358 about). Buridan was a French philosopher rector of the University of Paris who, it is said, described an imaginary donkey which, in front of

Consistency and truth in nature.

two identical haystacks placed at the same distance from itself, remained motionless without choosing between either of the two and, precisely because it would have both wanted in equal measure since they were the same for him, he could not decide so that in the end he died of hunger. This particular example is nowhere described in Buridan's writings but took its name, although versions of the same problem go back at least to Aristotle (see De Caelo 295b32). The statement that describes complete immobility could be refuted for some variables due to the absence of a total equivalence and an actual simultaneity of two stimuli, as could occur in a complex animal valuation. The statement of complete immobility of the donkey in the world, as we know it, can undoubtedly be refuted but could, theoretically, be defended in the hypothesis of total objectivity of nature. A real rebuttal action would be challenging today if we could replace the donkey with a computer that, by analogy, we can call 'Buridan's computer.' We could instruct a computer to choose for the best alternative (in the sense of a quantitatively greater one between two different possibilities or characteristics of volume, weight, or other factors) and, in the presence of a total equivalence between two inputs, the computer can get stuck in the impossibility of deciding on the issue posed by absolute equality. This imaginary situation of complete equivalency between some alternatives could also occur in humans only by hypothesizing a totally and mechanically deterministic world. A total equivalence of evaluation, however, would be impossible in a human world that we know in which freedom of choice operates since man could voluntarily choose to choose for other possibilities, independently of the factual conditions present, even if these are not to him convenient

Consistency and truth in nature.

considering the actual and various material factors present according to as judged by outside observers.

MATERIAL COMPOSITION OF LIVING BEINGS

«To understand and describe a living reality, one always tries first of all to extract life from it; then you have your hand full of inert fragments, which unfortunately lacks only the link of life. Chemistry gives it the name 'encheiresinnaturae': (manipulation of nature) makes fun of itself and doesn't even notice it. "

(Mephistopheles addressed to a young university freshman, in Goethe's Faust, vv. 1936-41)

The problem of the material composition of living beings has seen the history of scientific thought separate between those who supported the derivability of living organisms from matter alone and between those who, in addition to the material bodily support, also saw a vivifying intervention by energy, categorically different from those known, to act on a simple matter, otherwise inert.

Alchemy, the progenitor of modern chemistry, had dealt with metallurgical technology only for empirically known reactions until the fifteenth century, lacking an actual theoretical knowledge of the chemical-physical laws underlying such chemical reactions. In relatively modern times, the wonder of the living world directed the thought and attention of scholars to the hypothesis about effective relationships between specific material structures and the various functions of living organisms is foreseen.

History has seen various theories on the material composition of living beings, sometimes influenced by unscientific principles, where scientists' prevalent view was the existence of an immaterial force called phlogiston. This theory was developed by the German chemist Georg Ernst Stahl (1660 -1734), for whom every combustible substance has a common factor, baptized with the name of phlogiston (from the Greek word phlox meaning flame). Phlogiston, according to Stahl, gives compounds the ability to ignite themselves in an evident visible form of movement such as fire. Coal, alcohol, and wood were, in that theory, considered formed almost exclusively from phlogiston because they are highly flammable matter. Water, salt, sand, and every other animal, vegetable, or mineral was supposed to contain the principle of flammability too, therefore, movements capacity, even if in different quantities.

The theory of Iatrochemistry presented, differently from phlogiston theory, a current of thought intended to clarify life directly as a function achieved through the sole composition of elements ordinarily present in nature, without further intervention of occult forces. The living world was so described in the first half of the sixteenth century by the Swiss doctor Paracelsus as a regular branch of chemistry and medicine: in this theory, the characteristics of the natural elements and their combinations, together with their structure, were sufficient to explain the production of all the phenomena that are attributed to living matter without any exception. Paracelsus, nevertheless, admitted the particularity for the human body of a natural, intrinsic, and specific spirit in the composition of its constituent elements. According to this conception, all other living beings were efficiently illustrated by the vivid analogy of the image of a chemical furnace, where life itself is nothing more than a series of chemical processes to be explained and eventually put in the relationship between diseases and their remedies.

In the twentieth century, more elaborate theories appear, consistent with the scientific advances of the period. According to the idea of paleontologist Teihlard De Chardin, there is a tendency within the matter that leads to ever greater complexity and, at the same time, to increase one's consciousness. This view follows a conception of him called the law of complexity and consciousness, which, in all evidence, found its epilogue in the emergence of man at the peak of natural evolutionary processes. These theories were undoubtedly also the result of the influence received by De Chardin in his stays in the East, with effect mainly due to the Yoga doctrines, which see life as a progressive process of consciousness present in the universe for an unspecified purpose. However, the oriental principles mainly inspired by Buddhism did not suggest to him the idea of the tendency for every structure to an asceticism towards God, a tendency and progress instead present in his thought. That kind of progress, not present in Buddhist techniques in general because mainly oriented towards the realization and purification of an individual 'self.' So, they were to him presented with an incomprehensible subjective activity, of one single consciousness, without end and not even objective, in contrast with the idea of a type of progress that in his vision should be addressed to something universal and all-encompassing to be defined as real progress for each and all men. In the thought of De Chardin, the emergence of consciousness explains the tendency of matter instead to pass from the non-animated state to the increasingly evolved and complex one and which finally allows arriving at a true unification which is the fruit of a path towards God according to its principle that recite: 'everything that rises converges', for which there must be a real end to which everything converges.

Ernst Haeckel, very consistent with his materialist conception regarding the characteristics of living matter, hypothesized the emergence of life from the bosom of an exclusively material and eternal world, in which life appeared to unfold through repetitive cycles. Haeckel described the development process of an organism starting from the embryonic stage up to its final form, acquired at birth by the mother organism as a process where 'ontogenesis summarizes phylogeny': the living organism runs through in its current individual history, ontogenesis, all the stages of becoming of its ancestral history, summarized in all the evolutionary steps, from the most primitive to those closest to the final form and in any case in all the previous states ones, which constitute his phylogeny.

Material composition of living beings

Paradoxically, this theory seems to be of unconscious support for the ideas of Teihlard De Chardin, much less materialist than Haeckel, since they concern a kind of tendency of matter directed towards ever greater complexity. In all evidence, the more recent organism, which summarizes all the previous living forms in its evolutionary history, is undoubtedly more complex and with greater consciousness, at least when it is possible to speak of consciousness or self-consciousness.

The Biologist and Philosopher Hans Driesch, a pupil of Haeckel, although starting from mechanistic positions, then came to a theory, called vitalistic

philosophy, which admitted a power inherent in the matter, unexplainable in his historical moment, a force that was at the same time also an information principle. Driesch called this vital principle with the original Aristotelian term of 'entelechy,' which he then hypothesized was able to bring matter from a primitive and partly disorganized state to its specific final form.

Organic and inorganic chemistry

Organic and inorganic chemistry

A historical residue of chemical classification is shown that still defines organic a branch that belongs to basic chemistry and better defined with the name of compounds of carbon chemistry not necessarily belonging to living beings.

The distinction between inorganic and organic chemistry does not reflect a natural separation of laws or operational fields between two fundamentally different branches of chemistry. This distinction was historically formed by use and custom before discovering that chemical reactions, considered organic because scientists believed to be achievable only in living organisms, can occur according to all normal chemical reactions, which can be carried out in appropriate laboratory conditions.

Jöns Jacob Berzelius, in 1807 introduced the term 'organic chemistry' to define compounds deriving from the kingdom of living organisms, considered categorically different from those of the mineral kingdom. On the other hand, already at the beginning of the nineteenth century in the scientific community, the notion of 'life force' was a well-established paradigm, and Berzelius himself had determined its affirmation with his scientific authority. At that time, they believed that the synthesis of organic compounds needed a mysterious power exclusive to living organisms. Therefore, it was considered impossible to prepare organic compounds belonging specifically to living organisms in the laboratory because inorganic elements alone were devoid of the opportune 'vis vitalis' (vital force). Vitalism started from this notion, a doctrine that hypothesized a 'vital force' different from those normally operating in the inorganic world, and that was the first

81

attempt, apparently scientific, to respond to the problems that the nascent biochemistry posed and which, however, they seemed unsolvable with the then known natural physical forces.

Friedrich Woehler, a pupil of Berzelius himself, realized in 1828 the synthesis of urea, a chemical molecule then believed to be specifically organic because it is present only in living beings as a component of the urine of many animals, including man. The extracorporeal synthesis of this 'organic' molecule was carried out, inside a laboratory, by a synthesis between ammonia and cyanate molecules, and that is, 'inorganic' molecules. In this way, Woehler made a fatal blow to the various naive vitalist theories. In the following years, with the progress of the scientific disciplines, researchers canceled those ideas entirely from the list of accredited scientific theories. However, the elimination of the concept of life force had not resulted in the elimination of the term 'organic chemistry,' which is today no longer justified as truly distinctive of a different sector in the field of general chemistry. The representation "organic chemistry" today expresses a field that can be more appropriately defined as "carbon chemistry," which, moreover, can claim a branch of its own in chemistry in general, counting chemical compounds more numerous than that of compounds formed by all other chemical elements.

Due to the consolidated use of the term introduced historically by Berzelius, the scientific community continued to define organic as what was and is, precisely, belonging to a world of laws and compounds utterly pertinent to the field of general chemistry. Today in the conceptions of the modern chemist, the distinction between organic and inorganic conveniently disappears except for practical use to designate the chemistry of carbon which does not concern only living organisms for some specific privilege.

The formation of complex "organic" molecules certainly not present at very high temperatures may have occurred after cooling the earth to temperatures below 200-250 degrees Celsius, which happened in the Hadean eon between 4.6 and 4.0 billion years ago. If we consider the environmental differences at the earliest days of the earth: the extreme thermal excursions, of powerful ultraviolet rays, the electrostatic instability caused by the formation of violent electric discharges as well as the primitive miscellaneous chemical composition present in the oceans, we can understand the appearance of many and different compounds of carbon chemistry that, in the current environmental conditions, for most of them we can't witness a spontaneous formation.

In the early days of earth, after the first environmental cooling, about 3.7 billion years ago, began the formation of an atmosphere richer in oxygen, thanks to cyanobacteria action. Those microorganisms can use solar energy to convert carbon dioxide into oxygen. The increase in the amount of oxygen was then further increased by the subsequent and primitive terrestrial plant forms, and the lowering of ultraviolet irradiation, the significant reduction of ammonia, methane, and hydrogen cyanide then led to the formation of environmental conditions that eventually allowed life in primitive forms and those present in this period.

Today's atmospheric composition, with the relative lower energies involved, does not allow, as was the chance then, the concentration of reactive chemical species and activation energies sufficient to spontaneously produce appreciable quantities of the compounds of carbon chemistry as in the earliest days of the earth and therefore not even to those commonly present in living organisms. However, the biological organic compounds in current living organisms are produced because the enzymatic systems, which they are equipped with, can lower the activation energies essential for the various chemical reactions responsible for the formation of the biochemical compounds necessary for life.

Outside the scientific field, the use of the term 'organic compound' does not raise any problem if used in its proper meaning of historical etymological residue. Still, if used in an almost nineteenth-century way to agree with the idea that the formation of said compounds is the process for the construction of something belonging exclusively to the living world, then difficulties arise. The presence of organic compounds in an aqueous solution subjected to laboratory conditions with the simulation of environmental conditions, for atmosphere and oceans, similar to those described and present at the earliest days of the earth after four thousand million years ago, has led to the definition of 'primordial broth' also called prebiotic broth. This composition occurred following the formation of some combinations of chemical compounds, typical in such environmental conditions, that is, mixtures of organic compounds which, however, have been defined as if this 'broth' belonged to particular fields of chemistry, belonging to the quasi-living and not to absolute and scientific normality.

Even today, numerous scientific works use terms such as 'prebiotics' to define compounds and related chemical processes as if organisms were not formed by ordinary material components detectable on the face of the earth or in the atmosphere, and present in different combinations wherever there are certain environmental conditions. That without a necessary connection with the presence of a future life. From a certain point of view, almost everything is prebiotic. Only the structural organization of the physical and chemical elements considering their necessary functions within a living organism can make it possible to define a chemical compound as a prebiotic formula.

Unlike carbon compounds, the branch of biological chemistry deals explicitly with chemical compounds produced and used because they are found functional within cells or multicellular living plant or animal structures. Numerous compositions of carbon chemistry do not always have functions within living beings, just as for many inorganic compounds. For these molecules, the presence of specific functions both within the different metabolic pathways and in the composition of a living organism represents the crucial feature of fundamental importance for defining a given structure as really biological.

Structure and function

Structure and function

It is shown how the function is the final purpose of a structure differently from an aggregate that does not possess, in the strict sense, a function.

From a very general point of view, every existing thing has a structure when meant to have its spatial configuration. Language does not distinguish between structures that do not have a composition of the parts ordered to perform a function and between those that instead perform a function. Everything could therefore be thought of as a structure and can be analyzed indifferently, only relating to its composition, whatever it is. To better clarify the meaning of the material configuration of something, one should distinguish between the arrangement of the material elements of some things that are not organized for a purpose and therefore can be defined as simple aggregates from other objects in which the composition is arranged for a final pursuit and therefore can be defined as structures in a strict sense.

We can express the relationship between structure and function using Aristotelian terminology. The starting point is the sinolus (union) of matter and form since there is no matter without form and vice versa; they are one. It is not possible to separate them. The form is the organizing principle of the matter, not the matter itself. If a thing is what it is, it owes it to the form that organized the matter in that way. However, the matter must not be considered a generic matter, as the Philosopher states initially, with didactic intentions, but must be included in the form as a specific matter. A wooden sphere is not the same as a bronze sphere except for the shape understood simply in a purely aesthetic and graphic sense. A living being cannot be conceived as functioning with the same molecular arrangement: with silicon instead of carbon.

For Aristotle, therefore, the form is the functional unit of the parts. However, we must distinguish between a first and universal level order, which, common to all things, is conferred by the universal laws of nature with no function (function = 0). And the second level of order with a function reveals things that can be artificial or they can be living.

We will call things in the form with no function aggregates and forms that express function structures.

We can say that the final goal or function of a structure is the cause of the structure itself because it is only for the utility or function to be achieved that a particular structure, whether artificial or living, comes into existence.

To be more explicit, we have to say, regarding the concept of cause in general, that Aristotle conceives four types of causes that can be linked together, that is: a final cause that represents the goal to be achieved; a formal cause such as a project about the form to be implemented; an efficient cause that is the working action, and finally a material cause that represents the material means necessary for the accomplishment of what is accomplished or wanted to be accomplished. A classic example is given for a building that someone wants to realize: there will be a final cause as it will want to be built by someone with scope to make it; There will be a formal cause because, without a project, the shape of the building to be built would not be known by anyone, and so the same no strength could be addressed by anyone;

there must be several operators who will work as an efficient cause and finally, there must be bricks, cement, mortar and other elements available to compose the building, as a material cause materially. In this example, all four Aristotelian causes are evident. In the living organism, the final cause is represented by the final form that the organism must reach. About both artificial or living structures, failure to achieve their specific purpose degrades the functioning structure to an aggregate, in which there is no function.

One aggregate, for example, a rocky wall, can produce a particular echo when it reflects a sound. Still, the echo produced does not represent the purpose of the rocky wall, but, in all evidence, the sound produced represents only an effect deriving from the particular conformation of the rocky wall and not the purpose of the wall. The aggregate can be defined as such and distinguished from a structure since it lacks a final cause, at least if we think of a final purpose that is close to us and experienceable and not metaphysical as indicating a meaning in the totality, that is, in the known universal place of things in which everything makes some sense as a totality.

The presence of an ordered symmetry of the elements in material composition, due to particular chemical-physical conditions such as in crystals, indicates only and in any case an aggregate since it does not express any function.

Any relationship between material composition and a specific effect produced, such as the noise of a falling mobile phone, is such a relationship that it cannot be considered a real relationship, that is, a purposeful connection between a structure and its function as is the chance for the living or artificial structures. Has the cell phone broken?

Structure and function

It is not a structure in the strict sense, and because it does not work, it is considered an irritable and useless aggregate of the matter unless it is repaired. It is the function that gives meaning to any structure. Living organic systems are always organized for a purpose and always have their intrinsic final cause. In the living world, chemical molecules can be defined as structures endowed with function only within an organism that gives them this sense. Outside their position inside the organization of an organism, they can be defined as usual chemical species, even if belonging to the chemistry of carbon. This distinction will appear completely necessary when we want to determine the meaning better attributed to the concept of primordial broth and for what can then be defined as prebiotic soup. The statement that the purpose of a structure is the function remains completely evident regarding artificial 'structures' since they have been built by men evidently for some purpose. A problem arises when a system finds in itself, and for no other reason, the reasons for its development and the achievement of its final form so, as happens in living beings. The question then will be: how can a thing find its specific final form in itself when it exists only as an aggregate, that is, even before it exists as a structure organized to obtain its form? Indeed, the question leaves room for the idea of an external cause, that is, the idea of a creator God who gives a matter, present in a disordered and shapeless way, a functional organization of the parts dedicated to a purpose. On the contrary, scientific thought is today mainly oriented to hypothesize a beginning, at the origin of living organisms, due to a random molecular combination that, in some way, eventually arrive in a disposition being structurally organized in such a way as to possess the functions proper to live. However, this type of approach to the problem requires an analysis, for one composition, of

Structure and function

what we can define as 'essential function' so that a material configuration can be represented as a living being.

Aristotle (De Anima, II, 412, a27-b1) described the term entelechy as the ability in organisms to find their final form in themselves, as happens in all living beings. This term means 'inside' and 'purpose, from greek en + telos. The philosopher, however, did not manage to explain the phenomenon but limited himself to describing a particular kind of change that differed from any other phenomena that always present changes as the passage from potency to act but only through the work of other substances already in progress. Regarding living organisms, on the other hand, the change to reach the final form appears to take place due to an intrinsic project for which they already have their laws in themselves to pass from potency to act, apparently without the need for the intervention of informative external forces. Of course, the philosopher was well aware of the passage of information that natural laws operated through the seed coming from an already existing organism towards the progeny organism. But what about the very start?

To explain the existence of the inanimate world, Aristotle could affix a theory based on the world's eternity. According to which, everything is moved by other things, that is, through existing forms that are related to each other by interacting changes. However, about the existence of living organisms, devoid of external elements capable of inducing and establishing a final form, he could not present this theory, since it does not exist the eternity of the seeds already in place, except supporting the Platonic philosophy regarding the existence of eternal ideal forms, models of everything outside the sensitive experience, in which all living forms also participate.

With an intentional, but perhaps useless, shift in meaning trying to eliminate any possible attempt to attribute to an external and divine causality to justify the self-development of living organisms, Colin S. Pittendrigh coined a neologism, another term called 'teleonomy,' to alternatively describe the finality present in living beings to the term entelechy. He wrote: "*While biologists were quick to say 'that a turtle came ashore and laid eggs, it must be said that these verbal scruples were not intended as a rejection of teleology, but were based on the erroneous view that efficiency of final causes is necessarily implicit in the simple description of a direct mechanism. ... the long confusion of biologists would be removed if all systems so direct were described with some other term, for example,' teleonomic, 'to emphasize that the recognition and description of the orientation towards an end do not carry a contribution to Aristotelian teleology as for an effective causal principle.* "[20] (underlined by the author). From this writing, it can be observed by a reader that the scrupulous fear, which occurred to want to exclude a divine purpose in animal behavior, belonged only to the author of the term in question. The Philosopher's intention does not appear to want to insert an effective causal, initial and subjective action in creating living beings. Still, he wanted to 'describe' and not explain the peculiar characteristics present in the development of living organisms. In Aristotle's philosophy, the fact of not presupposing or describing something in place, other than the seed, which externally provided the rules of development to the single living organism, seems instead to suggest the idea of a precise release from the iron rule that requires the presence of a causal chain starting from an uncaused first cause. From this, it does not seem necessary to coin neologisms to replace the Aristotelian term of 'entelechy,' which in itself is sufficient to indicate a causal action that is not necessarily linked to a divine principle.

However, Jaques Monod will take up the term 'teleonomy' in the context of the neo-Darwinian theory to describe the purpose of a law that operates in the mutations of every living organism and represents a simple product of objective natural selection. However, the very object of our examination is the enigma of the origin of life. In that case, biological evolution is quite a different topic, and the evolution theory is out of the question in many aspects because it has excellent and consolidated explanations. We can observe that natural selection certainly cannot take place on organisms that do not yet exist, as it happens before the birth of the first form of life. The change of the Aristotelian term of entelechy with the term teleonomy does not seem to change the terms of the problem very much. It is pretty clear that the first living organism already must have had, in itself, the reasons for its specific development, its inner finality, but that not obtained by natural selection because certainly not yet possible. A theory on the origin of life should therefore not start from the organism in place as already developed but rather from what precedes the organism itself, in other words, from what would be in potency and not realized in the act. That said, in Aristotelian terms.

Indeed, one could consider prebiotics, potentially biological all atoms, and any molecular compound in general since all life ultimately has a material basis. However, the term should be reserved only for molecules capable of storing information such as DNA or RNA chains and those used in physical composition or organism regulation such as proteins. However, even these last molecules must still be considered aggregates until they are found bound in their organic functional sense. These molecules can be regarded as biological structures only when organized to function within a living organism, therefore in some metabolic or structural process. The function represents, especially in the living, the goal of the structure: any biological apparatus that does not work is eliminated over time by evolution because, if it does not reach its scope, it can only be defined as an aggregate, i.e., a set of elements arranged randomly, not belonging to the functional whole of the parts and not economic in the management of the living organism.

For a random set of molecules, the hypothesis of the passage from the so-called primordial soup to the living world represents the example, perhaps the most interesting, of the path from the aggregate to the structure, from an absence of function to the presence of process for the final purpose. The thesis concerning a very initial formation of a living organism, reached by pure chance and consequently lack of planning, was strongly supported by J. Monod. He begins his text by examining the objects that in any case present themselves to our attention relating to their belonging to specific build categories that he distinguishes between natural and artificial, noting the actual border between artificial objects created by man and natural objects believed created by randomness. With this beginning, Monod intended to affirm a design absence and, therefore, a lack of purpose in the origin of the living natural world. He tries to demonstrate already, before the successive arguments, the idea of objectivity of nature, implying the absence of any creative and finalistic intelligence, since the structures of the living world are natural them are not artificially made but naturally

94

made. The problem of how life began, posed in these terms, makes it difficult to imagine that someone could seriously think of a nature guided both in its origin and in its evolution, almost step by step, by one external creative intelligence deemed as an original cause because living organisms are natural objects, and nobody sees external hands making them.

As already argued, objects can more usefully be divided between those that present an effective relationship between structure and function and those that have a mere connection between aggregate and effect. These relationships are based on the most general cause-effect relationships, not based on the superficial character of material composition static. This distinction eliminates Monod's ambiguities regarding apparently artificial material arrangements because they are endowed with regularity, such as crystals since such designs are therefore to be considered, as stated, banal regular aggregates endowed with symmetrical characteristics in the molecular arrangement.

In "The blind watchmaker" [21], Richard Dawkins tackles the problem of complexity from a purely probabilistic point of view, affirming the improbability by the complexity of 'any' structure for any composition regardless of relations or any possible function. He cites examples of aggregates such as planes, piles of debris, Mont Blanc, and any biological structure as examples of equal complexity. It appears extremely difficult to establish beforehand what kind of probability they can obtain about their own composition and not another one. Indeed, he admits that biological structures show specific functions, but that does not mean that they are more complex. It only hindsight makes them judge more complicated for someone, given the wonder they arouse in us with their disparate and surprising functions.

From a certain point of view, nobody can deny that all things, even the simplest, are complex, at least when one has an interest in knowing and memorizing, for some reason, all the molecular configurations of all possible objects. We perceive that even the most minor thing is very complex, as the physics of nuclear and sub-nuclear particles can show. Apparently, with the same shape and composition, two coins are entirely different from each other thanks to the complexity of the molecular and sub-nuclear structures that constitute them. Among other things, these constituent parts have a greater complexity that we can't see because they cannot be described with analogies taken from our macroscopic world but are better described as centers of force having a spatial position identifiable only as of the probability of being in one place rather than another. The electron, for example, is depicted as a probability cloud around the nucleus of the atom to which it belongs. The complexity, artificiality, or naturalness of a material composition do not denote, for any object, a real ontological leap of state, but what permits this leap, in the way of being different things, is only being ordered for a function. The investigation of the origin of living structures, starting from fortuitous molecular combinations, requires the possibility of verifying the existence of any purposes to be achieved for given material composition. Only the presence of an end or function can allow us to judge one thing, whether a living organism, artificial structure, or the various concomitants of the elements of something regardless of its complexity. In dealing with the issue of the origin of life at the earliest days of the earth, the initial problem to be faced does not consist in the fact whether the synthesis of

Structure and function

organic compounds in general, albeit of considerable complexity, were possible, but how many of these can be defined functional and therefore to have sense in the metabolism or the structural organization of a living being even if elementary.

From the aggregate to the living structure

From the aggregate to the living structure

A reflection on the hypotheses regarding the possibility that random combinations of inorganic aggregates can acquire functions suitable for the formation of a living structure.

The transition from the non-living inorganic world to the living organic world can only be possible when imagined being accomplished through a structural change meant to the molecular level. That is, as the passage from a non-finalized composition of molecules, called aggregate, towards the realization of a design aimed at the life of a set of molecules which, due to the possessing a purpose, can then be defined as a structure in the strict sense. The establishment of molecular structures equipped with programs with processes that dictate the steps necessary for functioning, development, and self-replication are, in fact, the only real explanation of the origin of living beings. J. Monod liquidates the problem of the birth of life in a passage of his work and as: *"... The formation, starting from these substances, (nucleotides and amino acids) of the first macromolecules capable of replication"*. Pg. 114. Work cited.1

Except for mythological tales, the hypotheses formulated to explain the emergence of life on earth are, in general terms, attributable to only two: the abiogenesis achieved by pure chance and life creation realized by an intelligent entity. Thomas Henry Huxley used 'abiogenesis' to indicate the generation of life from non-animated matter and defined biogenesis as the generation of living beings from already animate matter.

From the aggregate to the living structure

In the context of the hypotheses that postulate the origin of life by abiogenesis, there is the theory of spontaneous generation which refers to the widespread belief in antiquity up to the seventeenth century, that life could be born in a 'spontaneous' way from inanimate natural elements under a specific life force. In modern times, among these theories, one can also place the idea of the origin of life by generation from the absolute randomness supported by J. Monod, in whose hypothesis the chance would replace the concept of life force as the first cause. The idea of spontaneous generation was challenged first by Francesco Redi and then again by Louis Pasteur in 1864, who demonstrated that life could not be generated spontaneously in an area free from contamination by other living beings: precisely as it would be by abiogenesis. The results from Pasteur's experiments were summarized in the Latin motto 'Omne vivum ex vivo' (no life is born without an antecedent life). In his experiment, Pasteur denied the possibility of a spontaneous generation had been, but those tests were carried out, necessarily for short periods, compared to those typical of geological eras. The experiments, however, could not refute the generation of life by absolute chance, which does not presuppose the passing of short times to originate living structures.

On the other hand, long-time observation was out of the tests' goal. So, if present, the presumed life force still had to generate life even from inorganic substances. The theory of abiogenesis due to pure chance instead presupposes a probabilistic, linked to the repetition of numerous events without any purpose, with the possibility of success over indeterminate times, almost certainly very long and in any case, for the hypotheses formulated, undoubtedly not due to presumed vital forces.

From the aggregate to the living structure

The hypothesis of biogenesis, in which every living form comes from a previous one, is presented as a particular and concrete chance of the process called 'scire per causas' (knowing through causes). Aristotle argued that knowing through causes, without placing a first cause which is itself not caused, does not lead to any knowledge since without this sort of the first reference, would occur an infinite series of reasons without reaching actual knowledge since all knowledge would always be relative to another precedent that must, in turn, to be known. Translating the concept, we can say that if a living organism is always the cause of another living organism, to find an explanation for living organisms, it would be necessary to postulate a residing organism in turn not generated by another living organism: a living not caused and since 'Omne vivum ex vivo, 'this first organism would be the foundation of all living things. From this, it follows that the hypothesis on the origin of life by biogenesis repeats in some ways an epistemological problem that cannot be solved empirically and does not eliminate the fundamental dilemma on the issue of the origin of life.

The hypothesis of biogenesis by creation, to have its credibility, requires to have a coherent, logical concatenation that descends from the postulate of faith of the design by God of the firsts or of a first living organism, from which would take place the reproduction of the others living organisms during evolution. From this perspective, chance can act as the change engine by selecting random genetic mutations only after this origin. In this hypothesis, we can postulate only the presence of a deterministic chance, by its nature capable of resulting in foreseeing determinable events—a theory which, however, does not exclude the intervention of a higher intelligence body.

From the aggregate to the living structure

The credibility of the hypothesis of the origin of life by abiogenesis, which should be realized through pure chance, requires a preliminary probabilistic analysis of the events of success, this latter represented by the realization of a combination of elements, capable of living, compared with the number of all other possible non-viable combinations. Today, compared to the past, such analyzes are possible thanks to the discoveries of molecular biology that have revealed the fundamental biological structures encoding the information necessary for the birth of life. This information is contained in specific molecules, organized primarily in polymers of different compositions and lengths. The research on the probability of emergence of the winning molecular configurations should be carried out on the simplest living structure imaginable, such as the one that presumably was the first on earth since it is undoubtedly the one with the most significant probability of coming true.

In the research conducted to discover the functions of a sort of biological molecules and the related molecular mechanisms, we always proceed to start from complex structures by carrying out operations of analysis and decomposition, that is, from the compound to the simple since this is the safest way and in many chances, the only possible way to proceed successfully in the search. Over time, these analyses have revealed more mechanisms and structures that indicate, even for elementary living organisms, a great complexity represented by the interaction of relatively primal elements such as nucleic acids and proteins in their sequences, combinations, and configurations.

The scientific discoveries that have occurred over the years have led to the sequencing of the human genome (to the discovery of the complete sequence of the nucleotide bases of DNA in humans) and the ever more in-depth understanding of the mechanisms related to replication, repair, transcription, translation, and regulation of the genetic code. Today there is no 'generically' a specialist in molecular biology, but many specialists deal with the various fields of investigation into which this branch of biology has progressively divided exponentially. Everyone can say this for any discipline, but this applies much more markedly than other scientific fields about molecular biology.

Regarding the origin and development of living organisms, the study of the information contained in DNA appears to be of fundamental importance. Because this can show the evolutionary history and transmigrations of human species through the aid provided by population genetics and bioinformatics, furthermore, these disciplines can also indicate the entire history of organ development in the living from a more analytical perspective over time than other archaeological research, from the most complex and modern to the simplest and most primitive of the past.

Analytically demonstrating the structure and functioning of the living machine helps understand the difficulties encountered in trying to reconstruct the opposite path, represented by the synthetic pathway, that is, the one followed, starting from the simple elements that make up the inorganic world, to arrive at the complex structures of the living world. The first step consists in initiating from the function to arrive at discovering the design that is, translated into the terms that affect molecular biology, to see how the synthesis of specific protein products is connected to particular DNA sequences and how this is made possible by the biological mechanisms involved. The advances in molecular biology have been enormous in this field, producing truly complicated techniques concerning molecular cloning, recombinant DNA techniques, and genetic engineering. These advances have also allowed the consequent formidable advances in proteomics, which represents the global study for research on the synthesis, structure, and function of proteins in general. At the moment, however, the synthetic way is not feasible, i.e., the ability to formulate nucleotide sequences, not copied from living structures, applicable to produce proteins and functional RNA sequences autonomously and initially according to an intentional design and for a specific purpose. What has been done, indeed, concerns the reproduction or cloning of arrangements pre-existing in different living organisms to produce proteins of great clinical interest, such as Insulin, Interferon, ATP (Tissue Plasminogen Activator), and many others, for that scope the genomic sequences revealed in humans and animals stored in bioinformatics banks are constantly increasing. Still, the codes of disclosed sequences do not make it clear how it is possible to go back from the knowledge of nucleotide sequences to the knowledge of the expression of a desired and functional biological product, except for experience in the field, that is, seeing what a particular nucleotide sequence produces in vitro. Today it is possible to synthesize DNA chains with the desired sequence. Still, if the series is not copied from an existing coding, it is challenging to randomly produce a valuable line for something. In reality, indeed, the researcher will probably find yourself making a protein with no usable functions.

Since the purpose of artificial or living structures, for them to exist as actual structures, is the function, we should be able to understand for organic molecules of any object of scientific research the specific tasks within the living being in which they are or should be founded. On the contrary, we could find ourselves faced only with aggregates devoid of functions and meaning. They are compounds, such that they can be found among the countless billions of possible molecular combinations that chance continually produces and destroys, but utterly useless from the biological point of view.

Among the hypotheses on the origin of life, a non-secondary place is a hypothesis that tries to answer the question concerning the possible presence of a connecting, intermediate structure as a link between the inorganic world and the living world. This issue arises because, among the forms of the natural world discovered up to now, there are no intermediate states that justify such a sudden leap in complexity between a simple mineral world and an incredibly more complex world, with the formation of functional and specific molecules of the living organisms. Since nature does not make leaps, there should be intermediate states in a progressive process to justify the existence of molecules such as mutually functional proteins and nucleic acids, entirely essential for life and its reproduction. Therefore, the problem that these hypotheses face concerns the research of what intermediate molecular compositions and configurations must exist to allow the construction of other more complex molecular structures, capable of finally making possible and explaining the complexity of living organisms.

Proteins represent the actual result towards which the processes of replication and development of organisms tend. However, in their isolated form, proteins cannot be taken as intermediates or as the real beginning of life on earth because proteins do not replicate themselves and are functional only within a living organism that should also be able to self-replicate with nucleic acid's aid. Scientific research around these issues has progressively been directed towards the so-called 'biological broths,' as primitive, prebiotic conditions, and on biological compounds eventually present in meteorites or planetary bodies.

The primordial prebiotic soup.

A look at the first scientific investigations to illustrate the formation in aqueous mixtures of non-elementary chemical elements at the earliest days of the earth and some considerations on the probability of formation of chains of amino acids, fundamental components of living beings.

Aleksandr Oparin in 1924 and John Haldane in 1929 independently hypothesized that in an environmental situation, as present on earth in its earliest days, a considerable number of various prebiotic chemical compounds could be synthesized. The characteristics necessary for these syntheses concerned an atmosphere poor in oxygen, rich in hydrides, an adequate supply of energy for the strike of lightning, the presence of a potent intensity of irradiation from ultraviolet rays. Such a scenario would have been possible in the primitive environmental conditions existing in the newly cooled and consolidated earth, along with the formation of the first oceans about four billion years ago. Some theories had hypothesized that the appearance of the gaseous mantle surrounding the ground was due to an inheritance of the solar nebula. In contrast, other ideas hypothesized the degassing of the liquid and solid components of the cooled and volcanic earth's surface. In any case, the atmosphere was then rich in methane, nitrogen, ammonia, carbon dioxide, hydrogen cyanide but almost devoid of oxygen; like it is present today in planets such as Neptune, Jupiter, Uranus, and Saturn. These planets, quite possibly, have preserved the primitive atmosphere thanks to their mass and the absence of bacterial or plant flora capable of converting carbon dioxide into oxygen.

The primordial prebiotic soup.

In 1953, an experiment conducted by Stanley Miller,[22] caused a stir in the scientific community as it seemed to finally open the way to understanding the origin of life on earth from its very beginnings. The laboratory experiment conducted by Miller witnessed the formation of compounds of carbon chemistry, belonging to the category of compounds that is still called organic chemistry today for historical reasons. The mixture of compounds was later referred to as 'primordial broth' or prebiotic. With this experiment, Miller demonstrated the hypotheses of Oparin and Haldane regarding the occurrence of some chemical phenomena, such as those likely had happened in the environmental conditions present in the earliest days of the earth. In fact, in a closed laboratory system, under particular conditions, i.e., simulating the conditions considered similar to those present in the atmosphere in the freshly cooled primitive earth, relatively high-temperature conditions and exceptional energy levels were being created, also by electrical discharges produced to simulate lightning storms. For the experiment, the researcher added various elementary components such as hydrogen, ammonia, methane, and water vapor to the laboratory system. Numerous molecules belonging to the chemistry of carbon were thus synthesized, including some amino acids, which are the elementary elements of proteins. The results of these experiments aroused enormous interest in the scientific world and society in general, interested in the problem of the origin of life, appearing to many as the demonstration of a way for the autonomous and spontaneous generation of life starting from the simple elementary inorganic mineral world.

Traces of an imine have recently been found in interstellar matter, a compound of carbon chemistry that can be used to synthesize amino acids. ISM.[23] was reviewed on the INAF newsletter (National

The primordial prebiotic soup.

Institute of Astrophysics. Italy) on 19 Jun 2020 under the title "Organic molecule discovery in the interstellar medium." However, the interest aroused in the use of the term 'organic' substance is not justified at all since it cannot be associated with the satisfaction of a real solution regarding the origin of life: the alleged solution to the mystery of the beginning of life is improperly supported from a mere linguistic misunderstanding because, as already illustrated, the term "organic compound," although semantically suggestive, is in no way support to the hypothesis that the molecules found to belong to a category of really intermediate compounds between the inorganic world and organic world. Such molecules are still enormously far from the complexity of the molecules needed in the constitution of a living organism. The conclusion appears to be made possible only through an inappropriate and deviant historical use of the term 'organic,' in this use, almost as if to imagine a different form of Vitalism, instead of considering some molecules, thus rediscovered, simple aggregate without biological meaning.

In scientific thought, the problem concerning the formation of molecules, specifically of living organisms, arose historically after overcoming concepts and ideas linked to obscure vital forces. In general, thoughts have thus emerged for the need to find demonstrations on the first origins of living organisms only through the aid of the laws of chemistry and physics, without resorting to incomprehensible forces and therefore abandoning any other reference of predetermined paths or predetermined harmonies.

Nowadays, numerous researchers have demonstrated the possibility of synthesizing organic compounds, even very complex ones

The primordial prebiotic soup.

such as polypeptide chains and nucleic acids, starting from different simple inorganic compounds. The research has shown that the organic compounds, which can be obtained, in various and controlled experimental situations, are now computable in several billion and obtainable through the ordinary laws of chemistry. Today, with the aid of some scientific instruments, many organic chemical synthesis experiments can be conducted parallel in different reaction chambers, varying the distinct reaction conditions for each compartment for parameters such as variations in temperature, pressure, and irradiation excursions. In the various experiments conducted, it immediately became evident that the number of different chemical compounds possible and obtainable is truly incredible regarding the variety of the relative molecular combinations, brute formulas, and the various structural arrangements allowed by stereochemistry.

In complex molecules, the variety of composition, compared to simpler molecules, is considerably increased since the brute chemical formulas, that is, the simple formats expressed in quantitative terms of the chemical elements present and their proportion, do not reflect the complexity of molecule chemical structures. That description should also report their positions in terms of spatial structure beyond the linear sequence of atoms. For example, many compounds have an identical brute structure, but two different spatial arrangements called chiral, as mirrored to each other, called enantiomers. There are configurations of higher levels for very complex molecular arrangements up to the structures of some connected sub-units. Only left-handed (L) amino acids and right-handed sugars (D) are found in living organisms, while this does not happen in environmental conditions outside the body.

The primordial prebiotic soup.

In Miller and other researchers' experiments, the investigations led to the production of racemic compounds, or a mixture of the two enantiomer forms, doubling the number of possible combinations. Numerous other compounds of carbon chemistry produced in the natural physical conditions are not found in organisms, demonstrating the redundancy of chemical forms present in the environment regarding those usable and functional to living organic structures. There are no known reasons why natural amino acids should belong to the L series rather than the D series. The fundamental thing is that they all have the same configuration because only in this way can they assemble the different proteins with a well-defined structure. Indeed, only one amino acid D in a critical point of a protein is enough for it to assume an incorrect conformation and become totally or partially inactive.

It is helpful to calculate for the synthesis of proteins how many different complex molecules are possible by combining the individual amino acids to understand the orders of magnitude involved in the combinatorial of the enormous chemical laboratory represented by the earth at the beginning. An example of calculating the number of simple amino acid chains obtainable is given by Forsythe et al. in 2017 [24]. Forsythe conducted a complete simulation of the primitive environmental conditions in a more sophisticated way than Miller. They included ecological cycles related to night and day that affected the earth even then. The experiment was conducted, with a few repetitive processes of temperature excursions to simulate an alternation of day and night in models of shallow ponds, added with a mixture of only three different amino acids and glycolic acid in water. A series of successive drying and hydrations were produced in the reaction system. After a few cycles, it was possible to find the synthesis of different

The primordial prebiotic soup.

depsipeptide chains (polypeptide chains with a hydroxyl acid monomer instead of amino acid). The polymers produced are easily transformed into polypeptide chains through small changes in the bond thanks to other chemical elements. The experience, therefore, found the formation of polymeric compounds with parts of different sequences, up to seven amino acid monomers. Of course, the amino acids' position in the polymer's linear line changes the chemical and steric properties of the final product, which therefore has its own particular identity. The number of possible combinations was, by researchers, then reported in the experiment for the four monomers (three amino acids + ac. Glycolic) synthesized in polymers with reduced chains of only eight monomers, arranged in all possible combinations. Despite a few reagent elements (three amino acids out of twenty available in nature), calculating the theoretical number of the different possible sequences for the various polymer chains resulted in 16,384 different combinations obtainable. In the experience, practically limited in time, for the four treatment cycles received from the reaction mixture, 33 different depsipeptides of 8 monomers were produced, represented by a chain of 7 amino acids + 1 molecule of ac. Glycolic. As for all abiotic syntheses, mixed, racemic mixtures of D and L forms were produced in relative proportions for each format at 50% of the total.

In the same work, furthermore, was appropriately detected the enormous different number of proteins that, in nature or artificially, could be formed in the presence of the complete pool of amino acids composed of twenty distinct units. Them presumably all present, in a possible primordial biological broth.

The primordial prebiotic soup.

It is known that the entire coding and transformation apparatus, represented by the DNA genetic code present in the cells, serves exclusively to encode and produce all the different proteins that will be used to form the various living organic structures and regulate their functions. Therefore, in addressing a discourse on the origin of life, it is necessary to begin to analyze as a priority the relationships between the structure and function of the various proteins present within living organisms.

In evaluating the relationship between DNA and proteins, it is worth noting a vicious circle whereby proteins encoded by DNA, such as DNA polymerase, are essential to synthesize the DNA while DNA is vital to encode proteins as DNA polymerase itself. According to the central dogma of molecular biology, the path of protein synthesis starts from DNA, rule in some respects overcome, but not relatively this problem, so: DNA transcription $\rightarrow$ RNA translation $\rightarrow$ protein synthesis. Furthermore, other distinct proteins must be needed beyond DNA polymerase for DNA synthesis. There are exceptions in the path outlined regarding retroviruses that use RNA as a template. However, this does not alter the meaning of the speech.

Indeed, we can see the same vicious circle for RNA synthesis. Some of the same regulatory proteins encoded by DNA are needed to stimulate or inhibit the genetic code transcription necessary to produce large proteins, namely RNA polymerase, an enzyme used to synthesize RNA. Other proteins are then needed to form the RNA-based ribosomal structure, which is, in turn, involved in the translation of the message present in the DNA to assemble the final polypeptide or protein chain that genomic mechanisms will use for the synthesis of RNA.

The primordial prebiotic soup.

Investigations concerning how life began from a molecular point of view are, therefore, still today, engaged around whether the chicken or the egg was born first. This question is an argument that brings water to the creationist hypothesis, at least until it is possible to solve this vicious molecular short circuit convincingly.

Leaving this unsolvable problem open at first sight, it seems necessary to establish the essential characteristics of a living structure to imagine the most uncomplicated possible organism. Miller's experiment had the merit of conferring a confirmation of greater breadth and 'naturalness' on the historic discovery of Friedrich Woehler concerning the formation of urea from "inorganic" elements. Miller's experience reproduced conditions that were carried out by nature without the aid of a chemical laboratory, confirming the assumption that matter is one and not divisible according to classes of ontological diversity, constituted according to a single natural law.

The term primordial or prebiotic soup acquires its meaning only when relating to what can be defined as something living. Aristotle states in this regard in De Anima (II, 412b 10 ff.): 'Of the natural bodies others have life, others do not, for life I mean the fact of feeding on oneself, of increasing, of perishing,' and again: (the book I 403b 25),' The animated being seems to differ from the inanimate mainly in two characters: movement and sensation.'

In modern times, thanks to the discovery of the genetic code Ernst Mayr, in 1996, could define a living being as: 'A living organism is an entity subject to natural laws, the same ones that control the rest of the physical world, but all living organisms, including their parts, are also governed by a second source of causality: genetic programs. The

The primordial prebiotic soup.

absence or presence of genetic programs indicates the precise boundary between the inanimate and the living world '[25]. Indeed, for a living organism to define itself as such, it must have a program for the desired aims to be achieved and implemented. Regarding the creation of a living being, a program must also define the biochemical and physical structures of the organism that is the way of producing its physical constituents; the regulation systems; the spatial configuration between the parts; the times of constitution of the different organs of the body and, beyond that, also the way of replicating oneself. In a much more articulated way, another author, Gerald Karp [26], proposes the biological concept of a living organism, defined as an emerging peculiarity, according to some functional characteristics for which something to be considered living must possess the ability to:

1) Evolution: To evolve, '(be subject to mutations: Ed.)' Thereby being related to all other living organisms.

2) Order: Be structured.

3) Coding: Contain, within itself, the information and instructions that control and define its structure and function.

4) Regulation: To be able to autonomously maintain homeostasis.

5) Growth and development: Being able to grow independently.

6) Energy: Representing an open thermodynamic system, able to assimilate energy, store it, transform it, and transfer it to the environment.

On that should, maybe, be added: Irritability, Sensitivity, and Motility: Being able to autonomously respond to external stimuli

The primordial prebiotic soup.

These last and more complete and specific definitions imply that concerning its functions, a living being, even the simplest, must have at least a DNA or RNA structure to respect points 1 and 3, a polypeptide or protein configuration to respect point 4 (to compose at least one external containment membrane of the organism), should possess at least one protein to comply with point 5, and presumably three or four protein compounds to comply with point 6.

At the present state of knowledge, it is impossible to imagine a smaller number of elements for a living organism. Antibiotics, in fact, by altering or eliminating one of the functions described above, kill microorganisms except for viruses because they lack the primary conditions to be defined as actual living beings.

Life does not simply require a set of molecules but not organized for a purpose and, at a minimum, it instead requires different chains made up of at least two classes of biochemical compounds: chains of nucleic acids and chains of amino acids working synchronously altogether, and endowed with precise functions. These molecules can be the components and sub-components of a more complex structure whose purpose is to live and replicate. Therefore, a set of molecules can be defined as structural for the living organism only if it has a goal in the living organism itself; otherwise, it will simply remain a prebiotic or abiotic compound: a mere aggregate, when correctly viewed, not what can be defined as a living organism.

PROBABILITY AND BIOLOGICAL MOLECULES

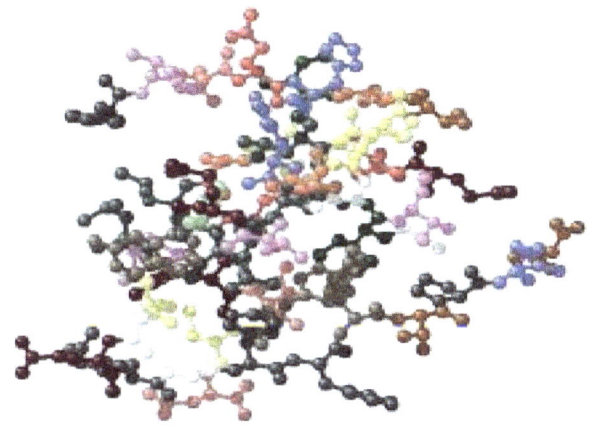

Insulin: a 51 amino acids protein molecule.

Each person makes conjectures, and on probabilities, he knows how to judge, reason, regulate his own acts and interests: this is, among our logical faculties, the one that is most constantly and essentially helpful to us in practical life. (Bruno De Finetti. Mathematician and Logician, on: Italian journal of statistics, economics, and finance. Bologna, 1933, year V, no. 4, page 723).

There is a link between the concept of chance and the concept of probability that Karl Popper expresses as follows: "The most important application of probability theory is the application to what we can call 'chance-like' or 'a pure chance' (random). They seem to be characterized by a particular kind of incalculability that disposes us to believe - after so many failed attempts - that in their chance, all known rational methods are doomed to fail. We have the feeling, so to speak, that only a prophet, and not a scientist, is able to foresee them. And yet it is precisely this incalculability that leads us to conclude that the calculation of probability can be applied to these events "[27] The classical definition of probability states that "The probability of an event is the ratio between the number of experiments in which it occurred and the total number of experiments performed under the same conditions, provided that this number is suitably large". According to this point of view, numerical probability assertions are understandable only if we can give a description or interpretation of them in terms of frequency: we cannot define a probability without considering a series of events and based exclusively on a single or a few experiences. A subjectivist definition states that: "the probability that someone attributes to the truth - or to the occurrence of a certain event, as a single univocally described and specified fact, is nothing more than the measure of the degree of confidence in its occurrence.

There are other definitions in this regard, but we can say that they are not mutually exclusive for what interests us here. To further characterize them, they can be divided into those that declare numerical formulations when expressed with numbers and non-numeric and subjectivist ones, as in the chance of J. Monod's formulation on the probabilities of the emergence of life. In it, he declares that it is like winning a billion in the lottery. Of course, a numerical expression of probability would come closest to the point of view defined as objectivistic, such as that expressed by Monod on the possibilities of the emergence of life.

To evaluate the reasons for own credibility, the hypothesis concerning one creative intelligence as the cause of the origin of life

requires only an act of faith in metaphysical forces that cannot be experimented empirically. The hypothesis that instead sees the creation of life by chance, in any case, requires a mathematical analysis of the probability of success for the occurrence of such a specific event. The research to be carried out for an initial evaluation concerns the mathematical relationship between a particular class of organic chemical compounds, which can be defined as 'biologically useful molecules,' towards molecules useless for living organisms as we know them. Naturally, this analysis considers the structure of functional molecules already operating in living organisms as a reference model. A good starting point can be represented by studying the mathematical relationship between a useful molecule model such as insulin, which has a few amino acids, towards all possible compounds of the same molecular order of magnitude in terms of the number of amino acids inside the polymer chain. This operation appeared helpful to examine the protein molecules class since they represent the final result of the essential molecular process for every living organism.

Only a few chains of amino acids or proteins in general, regarding the billions of possible sequences, have a function. An assessment can be made of the frequency of such an event to get an idea of the order of magnitude of the probability of obtaining proper molecules. Not all amino acid chains are defined as proteins. Only when these polymers exceed the length of 50 amino acid monomers are conventionally considered proteins; otherwise, with fewer amino acids, they are defined as polypeptides. Indeed, they are primarily hormones, transmitters, or regulators of biological activity. One of the small proteins of fifty monomers can be taken as a reference to evaluate the odds of randomly obtaining a specific single protein from a mixture of

Probability and biological molecules

amino acids. With the totality of all twenty different amino acids in nature available to make up this polypeptide chain, we can obtain with a simple combinatorial mathematic operation all the 'possible arrangements with repetition,' i.e., the number of all the different composable proteins, with the formula; D'n,k where n = 20 (the number of available amino acids) and k = 50 (The number of amino acids that make up the possible protein chain). From this calculation, $1,2589 \cdot 10^{65}$ arrangements of polymeric chains of amino acids are composable: the number of all composable proteins, different from each other and composed of 50 amino acids.

Nevertheless, if elucidation of the origins of life depends on a finicky reaction that occurs under narrow environmental conditions on an Earth-like planet only once in 1050 surface-mediated molecular interactions, then a detailed understanding of origins chemistry may be beyond current laboratory capabilities, even while life's origins is an inevitable feature of warm, wet terrestrial worlds.

A structure not designed 'a priori' but born by chance, as should be foreseen by the thesis of the objectivity of nature, requires the "blind" mode formation of the correlations between the elements that compose it. The realistic evaluation of the possibilities of a random composition of chains of nucleic acids or amino acids functional to a structure that can live requires knowing already at the beginning at least the probability of the necessary formation of a single first brick, to then multiply this probability to the other possibilities for other elements when involved in the connections of the biological building, imaginable as the first living organism. Success for not very probable events usually requires a long time because it is low the probability of success.

Probability and biological molecules

Assuming the rate of formation of one billion units per second for small proteins, such as those previously examined and, calculating the year as a compound of $3.1536 \cdot 10^7$ seconds, the probability of obtaining a single and chosen polymer molecule turns out to have a likelihood of success every $3,570 \cdot 10^{48}$ year or about three chances of success every trillion trillion trillion trillion billion years. Monod, in his book, does not say much in this regard, simply stating that a nucleotide sequence got the ability to pair with another in a complementary way.

The enormity of the numbers produced by this type of combinatorial analysis seems incredible. Still, it should be noted, for example, that the many usable characters ranging from 20 to 30 can form all the languages of the world while obeying the restrictions on the dispositions in words, of course, dictated by their own grammatical rules of each particular language, which limit some combinations also because they are often unpronounceable by the human phonetic apparatus. In Italian, in a simple four-letter permutation of the name of ROMA (Rome) "city," we can compose ARMO "voice of the verb to arm," RAMO, a term indicating "part of the tree," MORA "a fruit" or an adjective meaning a "woman with Mediterranean complexion," OMAR "personal name "this discarding the other 19 possible dispositions because they are meaningless. The example shown, differently from the composition of protein chains, refers to arrangements of non-repeatable elements of the set (that means that we can't put two letters of the same character in the word), as would happen, for example, for a word with maybe 2, 3 or 4 repeated identical letters, which is instead possible in the chance of polypeptide chains, in which it is possible to have even repeated sequences of similar amino acids, thus increasing the number of possible arrangements. In fact, with four amino acids for which we want to compose polypeptide chains of 4 elements, in a sequence with repetition as in the chance of protein synthesis, we will obtain 256 arrangements instead of 24 as in the example of the arrangement by permutation of the letters of the word ROMA. These formulations effectively explain the many possible protein chains starting from 20 amino acids.

But are the times offered to the chance infinite, or are there time limits to consider? The answers requested to these questions can be found in examining some theories on the universe's formation. On this topic, the most accredited hypothesis of the scientific world because astronomical observations validate it is the theory of the Big Bang which foresees a zero time, initial for the universe as we know it. The constatation of the progressive and accelerated distancing of celestial bodies between them has allowed us to ascertain, through the measurement of the degradation of the wavelengths of light, to place the beginning of the universe at about 13.7 billion years ago. ($1.37 \cdot 10^{10}$). According to the previous calculation, in this time, there would then be a probability of obtaining a specific polypeptide of 50 amino acids of about one in $2.60 \cdot 10^{38}$ Big Bang events, i.e., a likelihood of success of about 2.6 times every 100 billion of billion of billions of billions of times of the period represented by the length of time from the beginning of the big bang until today.

Of course, this calculation only serves to highlight the order of magnitude to be taken into consideration since an objection can be raised by some biologists about the need to select, as proper, only a single specific amino acid arrangement. Because there may be a certain number of protein molecules that possess the same functional meaning with somewhat different sequences, in practice, it could happen that not excessive changes in the polypeptide molecular sequence do not significantly limit the functionality due to a given protein. However, this is only possible to a limited extent since, in addition to the brute chemical sequence, the proteins must also respect a precise three-dimensional conformation, i.e., secondary, tertiary, and even quaternary structures (more complex degree configurations). These configurations are ensured only by an exact sequence and are not subject to extensive alternative changes.

The study of the molecular evolution of proteins has shown that there are many possibilities of having so-called homologous proteins for the same function, possessing similar sequences from species to species since they descend from a common ancestral ancestor. Some proteins such as cytochrome C are remarkably conserved in their line from species to species because they have few radically variable amino acid residues (Hypervariable sites), many sequences that accept similar amino acid substitutions for chemical characteristics (Conservative sites), and many sites are not subject to possible variations, (Invariant sites).

The analysis of nucleotide or amino acid sequences of the same or different species highlights the similarity in the amino acid sequences of other proteins that can be measured based on the percentage of identity between two or more aligned sequences. The rate of identity is nothing more than the fraction of identical amino acid residues in corresponding positions out of the total amino acids for the aligned sequences. From these techniques, it can be deduced that about 30% of amino acids in the linear structure of the protein chains cannot vary without losing their specific functionality. A practical example is provided by the cytochrome C mentioned above. This protein is made up of sequences of 102 or 104 amino acids for vertebrates and is thought to have appeared about 1.5 - 2 billion years ago. Since that time, this protein enzyme has not changed much in its sequence, despite being ubiquitous in many species. Some researchers have determined the sequence of cytochrome C in 100 different eukaryotic species, with an organism complexity ranging from that of yeast to that of man. From these experiences, it has been seen that the sequence of 38 amino acids of the 105 total is invariant.

In contrast, most of the remaining amino acids have been conservatively substituted. In only eight positions, 6 or 8 more different amino acids can be found for the same position proving to be sites hypervariable. The probability of a random production of a molecule such as cytochrome C, considering the essential sequences composed of 38 amino acids, can be calculated with the equation D'n, k where n = 20 (available amino acids) and k = 38 (the immovable dispositions in the amino acid sequence): The result expresses that the probability of success of obtaining that specific primary sequence, randomly, is one in $2,749 \cdot 10^{49}$. The probability of success to get this sequence is lowered by the structural constraints imposed by the conservative positions of the majority of the remaining sequences, which in any case approaches the probability of success for the proposed theoretical model regarding the random production of a short but specific protein of 50 amino acid residues.

Probability and biological molecules

It can also be observed, furthermore, according to what was stated, that some molecules are valid and operate only in specific sequences since they owe their function to a particular three-dimensional spatial conformation: hemoglobin is composed of four subunits, two by two, equal with three-dimensional conformation in the shape of a bag designed to contain and transport oxygen: these four subunits join together, and the sequence change in the subunits leads to some variant hemoglobin that is malfunctioning or often lethal when in the homozygous state. Immunoglobulins, too, must assume a precise three-dimensional structure to bind to the antigenic sites of the target pathogenic bodies. DNA polymerase III takes on a pincer shape that closes into a ring, a structural shape that is indispensable for sliding on the template strand to synthesize a copy of DNA. In any case, even approximating the number of polymeric arrangements for less than a billion possible structures does not alter the meaning of the discourse as a whole.

The amino acid sequences are also subject to evolution, and the proteins we know today did not suddenly arise, in their correct final form, in any case. In the example cited for cytochrome C, it has been seen that the current structures rest on at least 1.5 billion of previous protein evolution from perhaps simpler molecules, but for which the operative heart, that is, the set of invariable sites, can be defined as constant, or they could also come from other proteins with different functions and in any case, preserved because they are essential to the body. Useless and redundant molecules are invariably eliminated over time.

Probability and biological molecules

The enormous number of possible arrangements in the composition of a minimal protein chain, while seeming incredible if there were no limits of time and space, offered the possibility of chance encounters by the atomic and molecular elements, the apparent incredibility of the success of molecules biologically functional could vanish. These numbers, which represent almost zero probability of success in not infinite times, appear to be quite capable of offering good chances for successful events if an infinity of time occurs. A period in which eons and geological eras of the earth would be only a tiny part of an immeasurable period.

In time and space, an increase in the possibilities of lucky molecular combinations at the beginning of life could be offered by chance for a hypothesized phenomenon called panspermia through the opportunity that living matter is transported to earth from other places in the universe. The increased number of events on a universal scale would increase the probability of success for the 'birth of life' event. However, a lack of actual infinity of time in the universe, at least under constant physical conditions, appears to counter the idea of a significant increase in such probabilities since, for example, it is impossible to find molecules in some way organized to live at the moment of maximum contraction of matter, at the zero moments of the Big Bang and at some subsequent time, or similarly in the moments of full expansion. There are organized states of matter that are not suitable for life in such moments. Furthermore, in the hypothesis of vital forms originating on planets very many light-years away, the probabilities of a panspermia event are reduced due to the relative slowness of celestial bodies, which possess a speed not comparable to that of light. The panspermia that should occur due to the trillions of asteroids circulating in the universe and which could, in their impact, bring vital DNA or RNA molecules to earth has not found any credible confirmation in the findings of asteroid residues precipitated on the planet. In any case, any credible feedback would not significantly contribute to altering the order of the probabilistic numbers presented. The Big Bang, the common ancestor of the universe, is dated to about $13.7 \cdot 10^9$ years ago, the earth's rock formation is around $4.5 \cdot 10^9$ years, while the oldest-known star formed after the big bang is HE 1523 -0901 has an estimated age of $13.2 \cdot 10^9$ years, and so the formation of planets with optimal life conditions cannot be very distant in time. By comparison with the requirements of life that occurred on earth, since the two dates remain in the same relative orders of magnitude and do not have other planetary systems significant advantages for a longer time than the earth for a probable birth of life, considering the reduced speed of material bodies, the possibility of occurrence for the panspermia phenomenon appears to be strongly limited and improbable.

There have been many theories formulated on the life of the universe. Still, all have in common the affirmation of finite duration in time; therefore, a limited period is also due to considering the need for some physical and chemical environmental conditions suitable for life to start not present in however period. An exception is a hypothesis that would postulate a stationary universe, cited by 'F. Hoyle, et al. [29]. This hypothesis formulated in 1948 was developed as a secondary result of a theory concerning the continuous creation of matter capable of preserving the material density of the cosmos, which otherwise would decrease with the separation of celestial bodies from each other after the Big Bang. However, all the observations on the change of state of the universe have refuted this theory, instead affirmed the oscillating cyclical character of the universe while admitted, as a possibility, its temporal infinity in a continuing succession of contraction and expansion cycles.

In addition to investigating the time required to obtain a small protein, it is also necessary to examine how can replicate a particular protein. The probabilistic analysis performed so far focuses on the statistical possibility of obtaining the desired formation for a single protein that can be functional within a living organism. Today, we know that proteins for their replication in the biological field require the presence of information concerning their sequence, which must finally be encoded by pre-existing molecules represented by polynucleotide chains of DNA or RNA. This necessity, furthermore, make harder the chance of casual formation of an isolated specific protein.

The RNA world

The possibility of a single coding-translation system by a single molecular species is illustrated without the need for a duality of mutually operating molecules, as happens in current living organisms, between proteins and nucleic acids.

One of the most attractive attempts to answer the problems relating to the beginning of life, employing complex coding biological systems, is the theory presented in 1986 by the Nobel Prize Walter Gilbert for the so-called 'RNA World.' The hypothesis of the RNA world is a theory that proposes at the beginning of the earth forms of life, based exclusively on RNA molecules even before the formation of current living organisms, that are based above all on the functioning of biological systems requiring the coordinated functioning of DNA molecules, RNA and proteins, very effective indeed, but systems that present logical paradoxes to imagine their functional beginning. (see page 67)

Before 1980, the scientific community always believed that the enzymatic activities necessary for the body were performed exclusively by proteins. Still, the first ribozyme, namely a form of RNA with enzymatic capabilities, was identified in that year. In addition to this, the ability of RNA to self-model and replicate was also demonstrated, thus offering greater strength to the hypotheses that saw in this molecule the first unitary structural chemical form, perhaps capable of initiating the most primitive forms of life. One of the main enzymatic capacities linked to a particular ribozyme structure is present in the ribosomal sub-unit S23, which possesses the peptidyltransferase activities necessary for creating the peptide bond during protein synthesis.

According to the RNA world hypothesis, in cells, inside ribosomes and ribozymes, there would be a residual RNA form of the original RNA world representing a part of the primitive world from which the current biological system with nucleic acids would then evolve. Today's living organisms include RNA and DNA, which, compared to RNA alone, have considerable advantages in terms of stability and flexibility. Many objections have been raised to this theory because it has been observed that even in such systems, there would still be many unsolved difficulties in the formation of the unitary molecule of a possible RNA with dual enzymatic and coding functions. Some researchers then raised many doubts due to the significant greater biochemical instability of RNA compared to DNA and other factors common to DNA formation, namely: difficulty in obtaining the essential components for nucleotides present in the RNA molecule; the scarcity of a possible solution of available phosphates necessary to form the backbone of the nucleotide chain; difficulty in obtaining the cytosine and uracil bases in the simultaneous presence and optimal concentration. Such challenges would not suggest a real advantage for an RNA initiation over other molecules. Despite the criticisms addressed, today's most cultivated hypothesis remains the idea of the beginning of life triggered through a primitive RNA system. The main obstacle to accepting this theory always consists of the infinite improbability of obtaining the correct sequences suitable for forming truly functional molecules with a certain complexity, even in overcoming the inevitable difficulties of a simple biochemical formation without practical meaning.

The RNA world

The simplest and most direct way to analyze the possibility of an RNA world consists in examining the formation of an ancient and fundamental molecule such as is, for example, the ribosomal peptidyltransferase. This molecule is an enzyme called the common core of the ribosome: and that is, the nucleus of the center with activity for enzymatic protein synthesis. The sequence of this molecule is one of the most preserved structures present in the course of evolution up to humankind too. The tangible conservation sign is testified that changes in the complexity of this type of structure were made on modules represented by the common core only by adding and increasing RNA molecules and proteins, without altering the fundamental composition of the RNA chain. [30]

The RNA center so jealously preserved in the course of evolution, in its simplest form, has about 90 nucleotides which in its structure requires a very rigid sequence of four different nucleotides and which, unlike DNA, requires the Uridine in place of Thymine. The primary structure rigidity of that formulation is imposed by its absolute necessity. Because it is the operative heart, essential for proteins chains formation, that is the reason for its relative immutability of sequence during the phylogeny of organisms.

We can apply combinatorial calculus to evaluate the formation of RNA chains by the four different nucleotides to randomly make up the suitable molecule. The probability of setting up the correct nucleotide sequence must be obtained by the formula D'n, k where n = 4 (nucleotides available) and k = 90 (number of nucleotides that make up the chain); we can calculate a probability of $1 / 1,532 \cdot 10^{54}$. A random possibility of obtaining the proper sequence is entirely prohibitive.

The RNA world

Suppose we bet with an honest bookmaker on a horse with those odds of winning by betting a cent. In that case, we could have back, in the event of a semi-impossible win, a sum of around € 1,5 · 10^{52}. An unimaginable sum that does not exist on earth except for Scrooge McDuck with his billions. Furthermore, even admitting 1000 alternative, and in any case, functional forms of ribosomal RNA, would slightly alter the order of magnitude of the probability of success for the proper nucleotide sequence when obtained in a completely casual way. The hypothesis of the RNA world, which should explain or mitigate the exceedingly low probability of a random beginning of life by protein, dramatically lower the likelihood of a casual event favorable to life starting from the inorganic world.

Therefore, ribozyme formation before the genesis of proteins does not eliminate the problem of the random character of the synthesis of proteins. Naturally, in the absence of preordained plans, the presence of an RNA enzyme capable of lowering the activation energy for the formation of the peptide bond increases the synthesis rate of protein chains compared to an environmental production without pre-existing enzymatic systems. However, even this does not alter the chances of obtaining a protein randomly that is then functional within a possible living organism, except for reducing its execution times. The overall probability of success in getting the proper coupling of a correct RNA and then encoding the correct amino acid sequence is consequently further decreased regardless of how the peptide bond is formed. That is since it would be necessary to multiply the probability of obtaining the correct sequence for an RNA enzyme by the likelihood of getting the right arrangement for a protein enzyme capable of synthesizing the nucleic acid itself and some other protein. Indeed, RNA that codes only RNA, as we know seeing living systems, is pure nonsense.

The need to couple the probabilities of forming an RNA molecule with the construction of a protein molecule stems from the fact that no living organism can exist with a composition of only RNA / DNA or only proteins. To investigate how the simplest living organism could be, we should initially answer the question of the essential, minimum biological components that should have molecular structures to be defined as a living being. As stated in 1997 by Ernst Mayr (Page 58), a live design must have at least one RNA molecule for information storage and the possibility of replication and more, at least one protein molecule with enzymatic activity suitable for the RNA synthesis process. On the other hand, a protein cannot contain the information to produce itself and makes sense only as a helpful structure, operating within a living organism, capable of having the information for the replication of itself and the amino acid sequence of proteins.

A further requirement consists of the necessity of RNA molecules that encode specific proteins that can act as enzymes for synthesizing equally specific RNA molecules. A living structure, even the simplest possible one, not only has to produce 'any' RNA molecule or 'any' polypeptide chain, but there must be the correct coupled molecules.

Of course, the idea of such a simple organism represents an abstract model: because if RNA synthesis requires only one type of polymerase, unlike DNA synthesis, which uses three, in reality, the replication process generally also requires other different proteins.

The RNA world

Therefore, to support the idea of the creation of life, it is necessary to imagine the random production of RNA molecules and reciprocally functional proteins and their coexistence to interact effectively. The question, therefore, is: in a possible primordial broth, under optimal physical and chemical conditions, what are the probabilities that, by chance, RNA molecules endowed with a practical sense can be formed for the encoding of enzymatic and structural proteins, also functional to construct an elementary primordial organism?

One can imagine very high reaction rates to cause all possible reactions useful for the birth of life to occur randomly, but, currently, we come with difficulty conceiving the random formation of a form of RNA capable of self-replication. A replication, at the same time capable of producing polypeptide chains that, just as coincidentally, would be endowed with practical sense within an organism. The biologist must also consider those requirements for hypothetic organisms with absolute structure poverty.

In fact, by crossing the probability of obtaining by any case one of the functional forms of an essential protein, which we calculated to be one in $1/1.26 \cdot 10^{65}$ attempts, that number must be multiplied by the probability of obtaining a primitive RNA with a correlative meaning. This RNA is necessary for replication should have a simple sequence of 90 nucleotides, which we calculated to have a probability of formation by chance of $1/1,53 \cdot 10^{54}$ try. So, we obtain a possibility of success every $1 / 1,93 \cdot 10^{119}$ attempts. Even assuming a reaction rate of one billion of billion (10^{18}), for the formation of the molecules in question, we would always have a probability of success or the coexistence of the two desired molecules of the order of about $1/1,93 \cdot 10^{101}$ (sic!). Regarding the tiny possibility of a successful event, nothing changes in the probability's orders of magnitude, even if we consider a billion similar but equally functional shapes for the structure of each of the molecules.

When expressed in years to calculate the time needed to have a probability of at random formation of an encounter between the two specific molecules, we should make further operations. Following the calculations, we must divide the probability obtained as $1.93 / 10^{101}$ by the number of seconds of a formal year: $3.1536 \cdot 10^7$. Then, we would then get one possibility every 6.12×10^{93} years. Considering the big bang that took place (there is no absolute agreement) 1.37×10^{10} years ago, we can calculate, with an understandable approximation, a probability every very many billions of billions of times the duration in years from the big bang to today. Of course, as already stated, the numbers obtained should not be considered truly accurate to the billion but adequately express the order of magnitude for the probability of success of the expected events.

At the moment of current scientific knowledge, it is considered that a minimum structuring items that any living organism must foresee are: coding structures such as DNA or RNA, proteins useful to compose at least one containment membrane of the organism, proteins capable of assimilating energy, three or four protein compounds for the body material composition and, finally, the network of enzymatic proteins for interactions delegated to growth and development.

We will avoid the combinatorial mathematical formulation to calculate a result about the probability of obtaining, by chance, such a hypothetical and primary living being. For organisms endowed with all required elements, it would arrive at a challenging number to express, given the combinatorial analysis for two simple biological molecules.

These observations appear to be disputed in a scientific publication by Robert M. Hazen. In fact, in the conclusions of his work, Chance necessity and the origin of life, he writes: "*However, if the elucidation of the origins of life depends on a 'too particular' reaction that occurs in restricted environmental conditions a planet similar to the Earth only once out of 1050 surface-mediated molecular interactions, so a detailed understanding of the chemistry of origins may be beyond current laboratory capabilities, even if the origins of life are an unavoidable feature of the warm and humid terrestrial worlds.* " [28] The elements taken as the basis of his conclusions are the rate of reactions per second, (turnover frequency, TOF), in the formation of simple molecules of carbon chemistry such as amino acids, lipids, sugars, and nucleic acids. The time taken into account for his calculations includes a part of the Hadean and Eo-Archean period, about 600 million years, before forming the first elements of life represented by extremophilic bacteria. However, on these statistical considerations, the formation of life does not seem inevitable at all. The complexity in the formation of simple reactions of simple carbon chemistry cannot be related to the levels involved in forming functional proteins that make use of chains of amino acids and DNA or RNA molecules composed of chains of nucleic acids. These latter polymer molecules certainly use the availability of simple precursors, but polymers must be organized for a purpose. The nucleic and amino acid chains must collaborate to fulfill their replication function. The calculation for forming a simple protein of 50 amino acids is just a mere example. The formation and reproduction of life do not involve a single class of molecules. Still, they require a system of at least two different classes of molecular polymers which work together towards a goal according to a finalized project or, in other words, by entelechy. Living organisms require functional proteins that have a probability of random reproduction of 1.10^{165}, which must work in combination with functional nucleic acid chains that have the probability of at least 1.10^{54}. The formation of precursors of these molecules, even in the size of 10^{50} attempts, is a necessary but not sufficient condition to explain the origin of life in the 'short' period of 600 million years.

To overcome this obstacle cannot be invoked evolutionary mechanism, as often mistakenly occur. According to those explanations, organisms can proceed by successive and cumulative mutations from elementary structures to more complex systems, and that's correct. But as a crucial point to reiterate, evolution begins from a living organism and not before it. Despite the close to impossible and infinitesimal chances of success, the question, if ever it may be possible that chance can lead to life, remains valid. The answer can only be positive by believing in an infinite time available for the permanent roll of the die of life, which would thus run for eternity.

However, the most scientifically accredited hypotheses of the big bang predict a pulsating universe and exclude the idea of an infinitely stable universe. Instead, they were affirming a series, probably infinite, of dice rolls, which reset the results achieved at each cycle. As stated by the statistic laws, the success events for each die roll are independent and are therefore not affected by previous other rolls. In our chance, the fact that there was no life emergency for a million launches does not imply in any way that in subsequent launches, there will be more chances of success in giving rise to life, which always remain infinitesimal.

Despite the considerable difficulties encountered from a mathematical point of view, it seems that the statement that an infinite time can statistically produce anything, therefore even something like life, is now mostly peacefully accepted. However, suppose an alien traveler on a planet finds a switchblade. In that case, you can be sure that the traveler would not believe in a scarce and accidental statistical combination of different materials products. That is a much simpler structure to make than two biological molecules made for each other, even gifted with a self-replication program.

On a page where the problem of the birth of life is mentioned from the point of view of probability, not through mathematical terms, Jack Monod declares that modern science and biologists, as men of science, regarding a possible creative intelligence, must affirm: *"Modern science ignores all immanence. Destiny is written at the moment in which it is fulfilled and not before"* and, regarding the *almost zero probability of the emergence of life from chance, he declares 'Our number has come out on roulette: therefore why should we not feel the exceptionality of our condition, just like the one who just won a billion?'* [31]. To this consideration, we must observe that a single billion premium would represent a colossal scam, looking at the billions of billions of billions of a chance of winning. The fraud seems even more atrocious considering that the life so fortunately acquired should disappear at the end of a cycle of expansion and contraction of the universe.

Simple organisms?

Some single-celled organisms are described as simple with primitive characteristics that are not at all simple, as believed, but incredibly complicated. This makes it difficult to imagine intermediate structures between the mineral world and the first living organism.

In the past, lacking a detailed description of the molecular complexity and amino acid sequence of proteins in living organisms, it was possible to believe in Darwin's time that a few simple, randomly produced proteins were enough to give rise to life. In fact, in a letter to Joseph Dalton Hooker on February 1, 1871, Charles Darwin, anticipating the idea of primordial broth, suggested that the initial spark of life could have occurred in a: "small and warm pond, containing ammonia and phosphoric salts, light, heat, electricity, etc., so that a protein could be chemically produced ready to undergo new and more complex changes." He went on to explain that: "today such matter would be instantly devoured or absorbed, today such matter would be instantly devoured or absorbed, which would not have happened before the formation of living creatures." Darwin, therefore, claimed that existing life, feeding on simple organic compounds, could avoid the next spontaneous generation. Darwin's genius imagined how life on earth could have formed primordial, mixed elements well before Oparin's hypotheses and Miller's experiments.

At the very start, It is necessary to hypothesize an organism with essential and fundamental biological structures since it is presumed that the most primitive living being is also the simplest in terms of composition and functional organization of the parts. It is assumed that this organism, endowed with a primitive ancestral organization, may

Simple organisms?

have become extinct in the simplest way of single-celled evolution, supplanted by more sophisticated and competitive microorganisms. Today, we can take a primitive prokaryote of the Archea domain as an initial model of comparison, such as Nanoarcheum Equitans. This prokaryote is a single-celled marine organism, discovered in 2002 in a geyser on the coasts of Iceland, which possesses the smallest known genome for an organism capable of autonomous replication. Unlike viruses that cannot replicate autonomously.

Despite having replication capabilities because it can encode a mechanism for processing and repairing information, the Nanoarcheum lacks genes for the biosynthesis of lipids, various cofactors, amino acids, or nucleotides. For this reason, he must live in symbiosis with another Archea microorganism, the Igninococcus, in a symbiosis that allows him to spare the synthesis of the genes necessary for an autotrophic life. Like many other archaea, This elementary microorganism is an extremophile that lives well at a temperature of 80 °C with a salinity of around 2%. His genome comprises 0.49 Mbp (490,000 nucleotide pairs) expressing 540 genes. Production of proteins expressed by genes of 1090bp on average is then calculated, [32] from which we can assume the average for each polypeptide chain of a protein of about 200 amino acids. These proteins are assumed to be all functional for the survival and replication of the Nanoarcheum since few non-coding pseudogenes are present in the genome. Based on these data, we can believe that an even more primitive organism than the Nanoarcheum must have no less than 400-500 protein molecules.

The evolution of living organisms is a phenomenon that is wholly accepted today so that, for many biologists, the temptation has arisen to affirm that the life we find even in an essential organism is simply the result of an evolutionary process. But a pertinent demand should be: "evolution from what ?" The hypotheses relating to the emergence of life from the inorganic world should consider this problem, given the complexity inherent in a simple organism and the leap in complexity that separates it from the mineral world. On the other hand, the evolutionary process of the living world is out of the question, and it is, if anything, essential to know how evolution occurs in the entire molecular details, which are not yet fully defined is fully ascertainable. It does not seem possible to say in a generic way that life was born from absolute chance and then pause to describe only the subsequent evolution, which is precisely well known today and readily ascertainable.

The difficulties on the problem of the origin of life, strange as it may seem, are often dismissed as a problem to be overcome simply by talking about the evolutionary process. On the other hand, an actual explanation is needed because various hypotheses are formulated to solve the problem of the origin of life from evolution.

They are an evolutionary history of an organism that does not exist but pre-exists such evolution. An organism, then all to be explained. In fact, for example, We read on Wikipedia Italy.-it.wikipedia.org/wiki/Storia della Terra: 'Chemical reactions led to the formation of organic molecules that interacted to form even more elaborate and complex structures, and finally gave rise to molecules able to reproduce copies of themselves. This ability significantly boosted evolution and led to the creation of life'. This description is entirely correct, except that only what is already living evolves, not molecules. The result of life must precede evolution. On the other hand, what interests us is not given by what was created because we have it under our eyes every day, but above all by how and when the created life emerged. It thus appears necessary to get out of a purely descriptive framework of a given fact to enter into the merits of the life origin with an actual explanatory path.

The title of Richard Dawkins' book "The Blind Watchmaker: Why the Evidence of Evolution Reveals a Universe Without Design" reveals an awareness of this type of error. For Dawkins, too, it would seem evident that creation and evolution are different things since evolution presupposes a living being that evolves. In contrast, aggregates of matter do not evolve but combine. This distinction shows the weak side of some explanations and the difficulties already illustrated in talking about combinatorial mathematics in finding meaningful structural, chemical arrangements. The arguments of Alexandr Oparin and John Haldane, together with the Miller experiment, too, had the great merit of directing attention to the problem concerning the transition from the mineral world to the living world.

Dawkins' book takes its cue for its beginning from another text: "natural theology or, evidence of the existence of the attributes of divinity derived from the appearances of nature" by theologian William Paley. In this text, Paley takes the imaginary episode of a man who accidentally finds a watch as an example for his reasonings. Indeed, he concludes, an instrument of such complexity cannot be derived from chance but from an intelligent being, a watchmaker, as it would, by distant analogy, for God's creation of nature. Dawkins disputes this conclusion by stating that the works of living nature originate from simple, blind, and by chance. From this statement comes his explanation of the origin of life: he considers that origin and relative results as if a blind watchmaker built them. But, indeed, he loses sight of the fact that near the found clock, wanting to complete the analogy with a living organism, there should have also been other son watches. Was the blind clock so good at devising a clock that, just as blindly, it also came to possess a program for reproducing itself in order not to disappear from the evolutionary scene? Otherwise, the imaginary man described by Paley would have been fortunate to have found that one watch that appears, for a short peek, onto the natural world scene.

Dawkins, in his writing, tries to fill the gap concerning the discontinuity in the line that should show the progressive development of the aggregates from the simple mineral world to the first living being. The problem regarding this enterprise arises from the difficulties of finding such complex and functional structures in nature that can be considered intermediate between the simple mineral world and a first living being. Dawkins shows the impossibility, or rather the too much luck needed, to have a reasonable probability of obtaining complex and functional molecules of high molecular weight for proteins in living organisms by randomly mixing all the primary elements that form them. To imagine the formation of complex molecular structures as well as Dawkins does, it seems necessary to investigate the beginnings of life from the appearance of simpler molecules. The latter should be small proteins such as insulin and not hemoglobin, which appears very late in the evolutionary scene. To overcome this problem, he imagines a mineral evolution. A process that starts from simple elements of the inorganic world towards complex structures, like the evolution of living organisms. This operation, through small changes, will, by chance, become complicated, forming ever larger aggregates in a process that he calls 'cumulative selection.' In his opinion, this approach would lead to the formation of living structures in millions and millions of years of random constructions. Dawkins proposes simple examples for such an evolutionary beginning: those of the separation of small pebbles from large ones on a beach. That will be accomplished by the waves of the sea, in which he sees the creation of an automatically generated non-random order.

Another example is represented by a natural sieve that passes the smaller objects, separating them from the larger ones, as could happen for any geological crack. There would also be a creation of order in random processes starting from disordered situations. The motion of the planets and the balance between centrifugal and centripetal forces are unintentional too, yet the orbits of the planets are the most ordered one can imagine. With great intellectual honesty, Dawkins recognizes in the introduction of his book that he is a little oriented and the author of this book by his convictions. In any case, the great merit of these theories is that they do not deny a real problem with the origin of life from non-living things.

In analyzing the nature of what we call 'chance,' we have seen that there is no 'absolute chance' in the presence of natural laws which, necessarily, must carry the type of order illustrated by the theory of cumulative selection: in fact, the strength of the waves is more effective in moving small pebbles and therefore separating them from larger ones; the law of impenetrability of bodies prevents large bodies from passing through spaces and passages smaller than themselves; the law of gravity and kinematics allow the perfect "order" of celestial bodies. Beyond that, however, a genuinely orderly situation is not only a situation governed by laws but a situation in which there is also direct progress towards the realization of something. At the same time, nothing similar appears at all in the examples given. On the other hand, the natural forces, which should govern the accumulation of slight differences in the different aggregates joining, are part of the natural forces that equally lead to the disintegration of the same assemblages that aggregate. That could be a lack of neutrality of natural laws for, ultimately, favor the progress of a type of accumulation and difference in some direction. Analyzing the nature of what we call 'chance,' we have seen that there is no 'absolute chance' in the presence of natural laws. Those rules, necessarily, must carry the type of order illustrated by the theory of cumulative selection: in fact, the strength of the waves is more effective in moving small pebbles and therefore separating them from larger ones; the law of impenetrability of bodies prevents large bodies from passing through spaces and passages smaller than themselves; the law of gravity and kinematics allow the perfect "order" of celestial bodies. Beyond that, however, a genuinely orderly situation is not only a situation governed by laws but a situation in which there is also direct progress towards the realization of something. Nothing similar appears at all in the examples given.

On the other hand, the natural forces, which should govern the accumulation of slight differences in the different aggregates joining, are part of the natural forces that equally lead to the disintegration of the same assemblages that aggregate and therefore cannot be explained.

Simple organisms?

Why should there be a lack of neutrality of natural laws to ultimately favor the progress of a type of accumulation and difference in some direction. The objective neutrality of nature does not allow us to imagine progress channeled towards structures that are less stable than the original ones and require energy for their formation. These phenomena could happen when one imagines a project that directs nature, contrary to what the author of this theory of accumulation claims. The natural cycle of aggregations and disintegrations operating on sets of things does not lead to any real differentiation. Aggregates do not interest biologists interested only in the structures, really intermediate, ordered to an end, which is the birth of precursors of living organisms.

Even Dawkins does not avoid the logical trap, repeatedly pointed out, of using Darwin's theory out of its context. Darwin, of course, describes the evolution of living beings that change and evolve into other living organisms and not of minerals that structure themself to become living beings. In fact, in his book Dawkins writes: "An actual watchmaker has foreknowledge: he foresees the interconnections, having in view the future goal. Natural selection, the blind, unconscious automatic process discovered by Darwin and which we know today, explains the existence in every living being, the 'apparently' finalistic form. The man who came across the clock did not wonder if the clock (the living being) could evolve but rather who could have designed, built it or if it could have been born by chance. That is the question.

The neutrality of nature does not allow us to imagine progress channeled towards structures that are less stable than the original ones and require energy for their formation. Contrary to what the author of this theory of accumulation claims. The natural appearance of aggregations and disaggregations cycles does not lead to any real differentiation. This statement is correct also for sets of things endowed with a purposeless regularity. That does not interest biologists interested only in the structures, really intermediate, ordered to an end, which is the birth of precursors of living organisms.

Of course, even Dawkins does not avoid the logical trap, repeatedly pointed out, of using Darwin's theory out of its context, which, of course, describes the evolution of living beings that change and evolve.

A look on evolution

A LOOK ON EVOLUTION

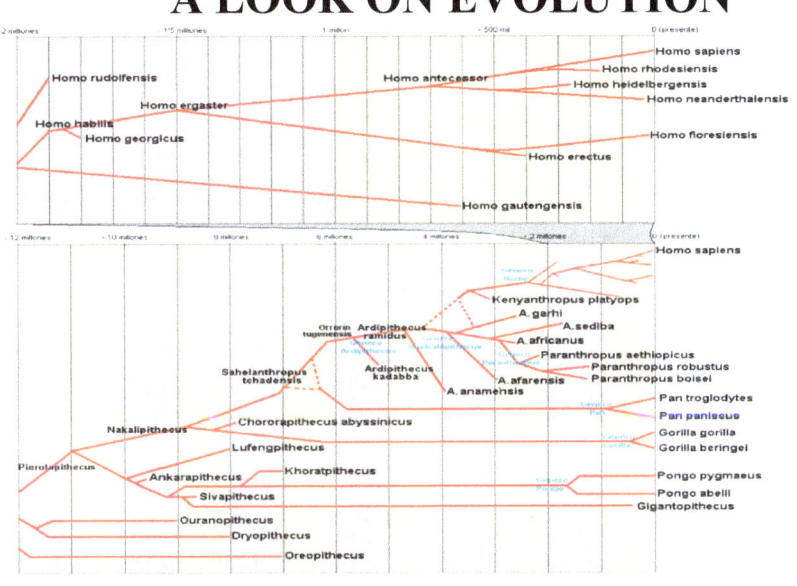

So, first it was Chaos, and then Gaia with her broad breast....

Esiodo opere: 1998 Einaudi-Gallimard; 2007 Mondadori, pag. 9)

One theory poses a hypothetical being, called in English 'last universal common ancestor (LUCA) or also "last universal ancestor" (LUA), or the 'last universal common ancestor,' as the first organism living. This emergence into life is estimated to have occurred between 3.5 and 4.0 billion years ago, and according to this idea, all current living organisms would descend from a very simple first ancestor. The oldest documentable forms of life are represented in the archaea kingdom, often defined as "living fossils," mostly polyextremophilic organisms, which are the simplest living organisms known. In contrast, traces of earlier and simpler organisms have not been found.

The beginning of life dates to about 3.7 billion ago [33], when the earth had cooled down enough to host the first living forms now known

149

after a billion years from its formation. I.e., the first archaea bacteria, which have left traces of their passage in the form of layered biofilms on rock formations. Before these findings, called stromatolites, no intermediate structure molecules were found that showed natural conjunction of continuity between the inorganic world and the living world. However, paleontologists could not find such paleo-chemical clues today because that world is too distant in time. Perhaps, they could constitute the first link in a food chain for more life forms that evolved later.

The evolution of living organisms has an explanation, today widely accepted by the majority of the academic world, in the work of random mutations for every living species and which occur in the reproduction of individuals. Not through changes in their organic conditions during their existence, such as supposed by Lamarck. So, the hypothesis of some directing intervention on natural selection by creators or orchestrators seems to be eliminated from the scientific horizon. In contrast to this point of view, however, another hypothesis does not deny the exact mechanisms of natural selection. That is the existence of mutations and selective conservation of the most favorable ones but supposes that these can be used or allowed for a purpose in the design of divine intelligence.

An actual antithesis between theories of life, based on absolute chance and creationist or dirigiste ones, does not arise, therefore regarding the problem of the evolutionary mechanisms of living beings but only for the passage from the inorganic world to the living world. In fact, what escapes out of our technical ability to control and forecast are typically considered as produced for pure chance. Hence, the need to propose sometimes a calculation of the probability of success for a given result, thereby abandoning the attempt to know the whole set of causes or initial conditions analytically. Creationist theories do not offer experimental scientific support, but they do not directly show a solution for those who support them. According to creationists, those theories can 'allow' us to glimpse an explanation wearing the glasses of faith, abandoning the useless attempt to know the whole set of causes or initial conditions analytically when impossible to see.

Unlike the hypotheses on the origin of life, the evolutionary theories must have as their starting object, for every possible explanation, a living organism from which subsequent organisms have been generated. This idea does not exclude the possibility of different lines of evolution, not starting from a single organism but also from other different organisms that originated independently of each other in different eras. In this chance, the statement 'Omne vivum ex vivo' is correct and corresponds to an evident logical necessity. A strictly empirical point of view recognizes that life, as we know it through fossil records and organisms still present on earth today. But that perspective does not give any evidence for elementary and intermediate organisms before viruses, which by the way, are not living organisms. Despite their simplicity, viruses must be considered successive in evolution to the bacteria, as they cannot reproduce except for cellular parasitism. They require, before that, a living organism in the strict sense, complete with molecular structures capable of replication and with autonomous metabolism.

The mechanisms underlying the evolutionary process are explained through the theory of the Neo-Darwinist paradigm, now universally accepted, which implements the original Darwinian theory with the acquisitions of modern molecular genetics. The paradigm states that the emergence of variations in genetic transmission within living species, with selective conservation of the beneficial ones, occurs for better adaptation and replication of the individual organism in the environment in which it lives. Another point concerns the range of all possible variations on which the selection occurs. Darwin, in his time, absents the concept of a genetic mutation which was coined lately by Hugo De Vries in 1927, considered a factor of variation in the species the genetic transmission to the offspring of the modifications achieved from animals during their individual life. According to a theory formulated by Jean-Baptiste de Lamarck. For Darwin, this was confirmed by the variations for the domestication of animals made by man, which were well known to him, through the crossing between the types that came to generate offspring with the characteristics most desired by breeders. These characteristics, however, were already inscribed as potentials in the genetic code and filtered in the domestication of animals, for which subsequent generations inherited the desired attributes in a more latent or explicit measure.

Contrary to what is often believed, Darwin, although openly professing himself agnostic in 1879, did not exclude the presence in the evolutionary process of an intelligent and infallible breeder. He borrowed that idea from many breeders' knowledge of selections for domestication. So he assumed and wrote. *'a Being endowed with sufficient insight to perceive in the external and internal organization differences that are completely imperceptible to man, and with a foresight extended to future centuries to watch with infallible care and choose for every purpose among the offspring of an organism produced in the previous circumstances. Nor can I understand why it should not form a new breed (or more new races if it were to separate the original organism stock and work on multiple islands) adapted to new purposes. Since we suppose that his discernment, his foresight, and the perseverance of his ends, are incomparably greater than these same qualities in man, we can be sure that the beauty and complexity of the adaptations of the new races and their differences from the original stock are greater than in the man-made domestic races ... Having sufficient time (and assuming that no known law can oppose its operation), such a being can reasonably aim for almost any result ... Seeing what has succeeded in making through selection the man, who is blind and inconstant, in recent years, and in what crude way he has probably worked, without any systematic plan, during the last few millennia, it takes great impudence to set limits to what the supposed Being could achieve during the entire geological periods'.* [34]

The creationist theories that historically presented themselves, naively ignoring the temporal dimension, indeed also present in the biblical account of Genesis, had produced the so-called fixist theories. Those declaring a single creative act had originated all living species fixing characteristics of organisms, separated according to species and genus from the beginning of the world and forever. These theories, based on questionable convictions, excluding the possibility of processes of essential change in the boundaries between the various living species, saw the creatures completely predetermined according to an initial established project. Chance, in fixism, was utterly excluded in the living world without the possibility of natural evolution, without prejudice to the adaptability and variety produced in the individual living during their own defined existence.

A look on evolution

John Ray, a proponent of fixist theories at the time he wrote, seemed adequate to explain the living world and did not appear to contradict the experimental research of the time. However, Ray was an eminent naturalist with excellent abstract and bookish and testing knowledge. He traveled to continental Europe from 1663 to 1666, cultivating his naturalistic studies and collecting some remarkable specimens for different plant varieties. Despite the prevailing idea of the lack of evolution in nature, great intellectual honesty wrote: *'Some condemn the study of experimental philosophy as mere curiosity and denounce the passion for knowledge as an undertaking that displeases God. And so cooling the zeal of philosophers as if Almighty God was jealous of the knowledge acquired by man ... as if He regretted that man uses the intelligence that He has given him to study the objects of which he has enabled him to understand and which he has offered for his investigation '*[35]. Ray still wrote the works of creation are: *'works created by God in the beginning and preserved by him until this day in the state and condition in which they were made.'* [36].

Before John Ray, it appeared quite possible to overcome fixist theories even by Christian philosophers in the history of thought. Thomas Aquinas thought that the autonomy of nature was a sign of God's goodness and that there was no conflict between the concept of a divinely created universe and the idea that the universe could have evolved through natural mechanisms. Aquinas could not argue on how, in what way, this evolution could happen. However, even admitting a natural evolution, he opposed the idea of the supporters of Empedocles, who affirmed that the universe could have developed even without a final goal and only by chance: Commentary on 'De Anima.' [37]

154

A look on evolution

Through evolution, God's way of working is present in an embryonic way in the Age of Enlightenment. Immanuel Kant writes: *'Matter which is the original substrate of all things and therefore linked to certain laws, when freely abandoned to such laws must necessarily produce beautiful combinations. It has no freedom to depart from this perfect plan. Since, therefore, it is subject to the wisest design, it must necessarily have been placed in such harmonious relations with a First Cause, which dominates over it. This cause is a God, as precisely shown by the fact that even in chaos, nature cannot behave except with regularity and order.'*[38], (author's emphasis). Kant clearly states that even within the sphere of natural laws, matter 'abandoned' to such rules can produce beautiful combinations' even if necessary, following a final divine plan.

Neo-Darwinism rejected Lamarck's evolutionary theories as erroneous despite Darwin's acceptance of these hypotheses. The rejection of Lamarck's views, in later times, happened since he affirmed the inheritance of the characters and characteristics acquired through the use or disuse during the individual's life. Therefore, this mechanism was considered a kind of non-random direction by the individual organism on its genetic heritage. In contrast, the central mechanism of the neo-Darwinian evolutionary process is based on the phenomenon of completely random mutations in the genetic code. These changes occur in the reproductive cycle and are ultimately passed on to the offspring, but they are not, in their occurrence, directly connected with individual animal behaviours.

During evolution, the elimination of individuals who are found to be weaker takes place, that is, those who, due to their characteristics, are less suitable for surviving the existing environmental conditions. Only coincidentally, more suitable organisms at that moment survive and can transmit their characters to progenies. The variables on which Darwin's evolutionary theory focuses regarding the species formations are the mutation of characters, the inheritance of innate characters, adaptation to the environment, the struggle for survival, natural selection, and geographical isolation. Through the improvement of these points, individuals of a living population can compete with each other for natural resources. The idea of the struggle for survival, of which animals have no concept, was taken up by the thought of the economist Thomas Robert Malthus. At the beginning of the nineteenth century, he published: an 'Essay on the population principle.' He stated that the increase in the population for a specific available cultivable area was much more significant than the increase in food resources. The philosophy of Herbert Spencer included this idea in his theory of social Darwinism in man in analogy with the struggle for survival in the animal world.

In terms of physical factors and natural resources, the variability represents the discriminating filter for the affirmation of one species compared to others, thus defining the idea of natural selection. An important factor is also the geographic isolation of living populations. At the beginning of the geographical separation, the accumulation of the different variations produced in the genetically homogeneous groups, at the beginning of the geographical separation lead, through subsequent mutations, to profound genetic divergence until creating genetic makeup very different in the separate groups. So, eventually to make they do not inter-breeding and therefore become other species. In Darwin's theory, modified with the advent of modern genetics, regarding genetic reproductive succession, the idea of a genetic imprint of the individual's adaptive effort is then eliminated. In the adaptive success selected from so many possible mutations of the parental DNA, the concept of randomness is necessary to explain the evolutionary process that ultimately leads to the forms of organisms that are most suitable for the environment in which they live.

From the premises set by Darwin's theory of natural selection, many thinkers see the 'chance' as a cause of evolution, opposite and incompatible with the idea of divine creative intelligence. Despite being seen as entirely inevitable, the idea of a total opposition between chance and creation is not accurate since God could use the way of operating that requires the 'chance' to produce determinable outcomes. As we have seen, there is randomness that can be defined as necessitating and which can therefore be deterministic like any other cause-effect relationship. This way of operating and determining results through "chance" also occurs when dealing with entirely random initial conditions phenomena. It then results in a wholly predictable outcome as occurs in the evaluation of temperature, pressure and chemical balances: all physical situations in which randomness is present on numerous events or particles. In a natural evolution, randomness becomes necessitating and deterministic, for example, when mutations occur in populations with a certain number of specimens. For example, the rate of genetic mutations in a given population of living beings varies significantly from species to species for the various possible modifications. But, also in these cases, the chance becomes deterministic when considered for a long time, as occurs for geological periods. The mutations certainly happen randomly but express results determined by the number of mutations occurring in a specific time frame and for a particular species. In addition, the preserved modifications are either advantageous or indifferent to the adaptation of the mutated species. Still, they are not arbitrary and totally by chance but must have paths channeled into the possibilities imposed by natural laws.

The presence of "chance" in its deterministic meaning does not allow the doctrine of Neo-Darwinism, based on the absolute randomness of mutations, to cancel the idea of the possibility of creation and evolution of nature. Certainly not by miracles, but through the laws that God has created and personified as the main religions claim. Of course, saying: "that himself created" can mislead the mind to think of a God absolutely detached from the world instead of a conception of a God who, even if transcendent, as a cause, does not separate himself but remains in effects. That is, in creation. Conversely, the idea of a creator who mechanically determines creation and evolution in all the living nature comes into conflict with the dogma of human freedom by removing all moral responsibility from the man who, as can be seen, today, instead bears responsibility for many nature changes. The fact that nature can evolve through deterministic chance' removes the characteristic of rigid and mechanical determinism in unfolding natural events, as would happen for a first and direct cause without any mediation. Man can, therefore, freely intervene in biological processes and alter natural evolution, which then, through necessity and deterministic randomness, will find new paths that are not always predictable due to man and his free will. Of course, all the outcomes of all the possible ways to which natural evolution could arrive, and then, in fact, will come, cannot be unknown to a God who, by definition, is omnipresent, omnipotent, and is also the laws themselves.

From that point of view, what can alter the evolutionary paths described by Neo-Darwinism is the appearance of man in possession of free will. The situation just told somehow recalls Leibniz's distinction between the truths of reason, which can best be defined as 'entities' of reason, and the 'truths of fact.' In the presence of 'chance' as it is generally understood, dominated by natural physical laws and theoretically predictable in its results, we can call 'truths (entities) of reason' all potential events that are undoubtedly feasible in the future. These entities (truths) of reason, what Leibniz defines as infinite, are not truths in the strict sense but possible outcomes for mental operations that derive, by internal coherence, from the premises. So, from them, all the possible combinations of events can arise. These truths of reason are not the 'facts,' tangible entities. They are not facts but entities of reason. That is, conclusions that logically depend on the premises and which, mathematically and abstractly, are endowed with intrinsic necessity.

An excellent example of this can be provided by chess, card, or board games. In such games, not all moves are possible. Still, only some are feasible: that is, only those logically possible should be in the player's mind. The more possible events are in the player's memory, the better gambler he is; the more moves and potential events, the more is difficult the game. After each move (i.e., the truth of fact), the scenario of possible moves changes as a set of entities (or truths) of reason. That is to be memorized by the gambler. The number of possible events, as named the truth of reason, is predictable in any game situation or scenario. For these grounds, computers are better than humans at such games.

This game of possibilities can also be exemplified with fewer alternatives by the less complicated situation in which a man is found to jump a small precipice, chased by a tiger. He must choose between two possibilities 1; stop at the edge of the ridge and 2; jump low. Then other options will open: hypothesis 1a, the tiger, will be able to attack him; hypothesis 1b, the tiger wants to play; hypothesis 1c, the tiger, will chase another animal close to him; hypothesis 1d, the tiger is not particularly hungry and so bored goes away, etc. In the chance that man is opting for the jump, we have possibility 2. Then: 2a, he crashes to the ground and dies; hypothesis 2 b, he survives but some bones fracture; hypothesis 2c; he lands completely unscathed; hypothesis 2d, he manages to stop along the wall by clinging to some bushes; etc. Of course, one possible hypothesis 3 is discarded. For which we could have: 3a, the man relocates to a bar in the center city; 3b, he leaps and levitates and lands softly; 3c, he turns powerful and disintegrates the tiger with its hidden power; 3d, he jumps backward with a jump of three kilometres; etc.

In the game of life, the number of possible events is almost infinite, even if their number is less than that of impossible circumstances. As a factual truth, every event changes the number and overall scenario of possible events we can know as truths or entities of reason. Only a god can have in mind all the possibilities, as a truth of logic, or in other words, the hypothesis of possible events, to occur as factual truth. Of course, this is another way of expressing the passage from power to act that always applies to man. Indeed, if, in any system, an entity of infinite intelligence thoroughly knows all the events, which for man are considered potential, the character of potentiality would disappear in him since only one result from the many thought possible happens. And only a genuinely unpredictable component of the system can make all events equally unexpected and the premises genuinely potential.

Suppose the premises, consisting of all the natural conditions present at a given moment, are not subject to changes due to a real absolute chance, in the sense of total unpredictability, since the existing physical laws tightly determine the evolutionary process. In that case, the path of nature remains entirely determined by the iron cause-effect sequence with a single truth of reason which inevitably and necessarily becomes the truth of the fact. That is what happens.

With the consequent absolute unpredictability, absolute chance fits into nature with the appearance of man and his faculty of free choice. What finally happens depends on the possible human choices that are entirely unknowable. Thus envisaging, differently from a necessity and foreseeable randomness, in the absence of man, numerous possible evolutionary outlets linked to equally thinkable truths of the reason that only a God can know in their totality and that, of which, only one will become the truth of the fact. That is, what will happen.

A situation contingency then arises for the results achieved by the evolution due to variable and unpredictable premises, as can happen precisely for man's intervention. So, if the faculty of free choice is admitted for man, one must also admit that his freedom is connected with the possibility of freely managing nature. Which would not be possible in a direct decision-making mechanism by a divinity that eliminates any freedom and responsibility of man towards the environment in which he lives and works.

The evolution engine

Genetic mutations are described, their role in the evolution of living organisms, and the strength of the arguments supporting species adaptation in natural selection.

Darwin has shown that random mutations in organisms are the actual biological engine of evolution. His theory, however, has no predictive value regarding the future evolutionary path for the various living organisms. We can better understand the past natural evolution by considering how many species have died and not just how many have survived. Today there are at least eight million living species, and it is estimated that over ninety-nine percent of all species that lived in the past are now extinct. It seems to be a total lack of economic management on evolution phenomena, which can thus be seen as an occurrence that would exclude a plan intelligently preordained towards creating increasingly competitive and perfect forms. However, suppose evolution does not proceed by leaps or dint of miracles. In that case, it must proceed through natural laws which provide for random mutations and the survival of the most suitable organisms. It calls for organisms extinction of less biologically equipped in the competition for nutritional resources and the defense of the proliferation capacity of their particular species.

So, we can only ascertain evolution with hindsight, the development and affirmation of living species, and understand the survival strategies of the various organisms. Still, we cannot predict what the future winning mutations will be. The mechanism of natural selection states that a particular organism survives when best suited to the natural conditions of the moment, but this adds nothing to the understanding of why a specific species and not another gets a winning mutation and, therefore, why that very particular species survives. This consideration is also called an analytical judgment by pundit logicists. Indeed, everyone can demonstrate conclusions even with logical reasoning without empirical proof. That is not an actual demonstration but only a conclusion consistent with the initial assumptions that affirm that the most suitable to survive then survives. Biological structures have the primary survival function, so naturally, they are adapted for that. So, when viewed in terms of structure and function, some aspects of the theory of natural selection can be seen as a form of a tautology because, if a function is achieved effectively, it is pretty evident that the underlying structure is suitable with no wonder. This tautological form concerns every species and every evolutionary step and does not allow us to predict something in the future of natural evolution. The affirmation that the most suitable survives to continue living means that its evolutionary history has made it suitable to live and reproduce; this is verified by the fact that its mutations have been successful for the sole fact that the organism survives. There is no need for other explanations: if the organism lives, natural selection states with a retro-diction that, thanks to how it has changed and structured, it must live. Therefore, the theory does not arrive at predictions on the evolution and success of any particular organism or animal species but affirms, and in a nutshell, that what lives can survive because it is suitable for living: an argument that is difficult to dismantle.

The explanatory and tautological power of natural selection is compelling. Its stringent logical argument is also used in knowledge sciences, inspiring the theory called evolutionary epistemology. In this theory, applying the criterion of natural selection in the field of knowledge, it is postulated that not all the very many ideas and technical procedures produced by philosophical and scientific thought survive. Many perish, and only those considered most valid are retained since they resist refutation from other ideas, theories, and procedures. They, so, are preserved as an inheritance of acquired knowledge until better ideas supplant them. Thus states K. Popper: *'According to my proposal, what characterizes the empirical method is how it exposes to falsification, in every conceivable way, the system it has to control. Its purpose is not to save the life of unsustainable systems, but on the contrary, to choose the system that in comparison turns out to be the most suitable after having exposed them all to the most ferocious struggle for existence'.* [39]

The most immediate way to demonstrate the theory of natural selection is by experimenting on bacterial populations grown on soils containing an antibiotic and the necessary nutrients. Only bacteria that, quite by chance, acquire a genome mutation called the 'R' factor, which confers specific resistance to the antibiotic, can reproduce on the medium containing the antibiotic in question. Thanks to this experience, it is possible to predict that, after a certain period available and with a large population of bacteria, some individuals of the bacterial population will acquire resistance. This example also shows a problem, relatively widespread in hospitals, for which new antibiotics not "known" to bacterial populations must continually be developed by the pharmaceutical industry. The new drugs developed will remain helpful in the entire period in which resistance due to a genetic mutation in the bacterium to be combated has not yet developed.

The evolution engine

One problem has arisen from some philosophers about the epistemology status of darwin's theory. Is that a scientific one? The Falsification Principle, proposed by Karl Popper, is a way of restricting science from non-science. According to Popper's principle, natural selection theory is falsifiable, so it appears scientific. The falsification, if concluded positively, validates some theories as genuinely scientific. That because not only experimentally verifiable but also controllable and, in case, refutable for the possibility of practical or theoretical error. Suppose it was not possible to ascertain errors. In that case, anyone could question a statement since any statement will always be valid and not subject to verification. So, scientists should test every theory for the truth (or falsification) to see if it could be declared false about its claims through any imaginable or real check. According to Popper, religious doctrines are prototypes of theories that cannot be falsified and, therefore, not verifiable. It is impossible to imagine experiments that can contradict them. They are therefore certifiable only by faith but not as scientific theories. The theory of natural selection is falsifiable because it is possible to imagine an experiment or a situation in which it can be shown that the most suitable to survive does not always win, only for the simple fact that he lives. The survival of a less suitable organism against a more suitable one could result from an event that has favored the luckiest one who gets to be in the right place at the right time. For example, Some extinct species may have been disappeared for unfavorable circumstances or because, in casual situations, they did not survive the great massacres that our planet has gone through throughout its history. In this, we deduce that it is possible to empirically verify or hypothesize that, in some circumstances, determined species were not the most suitable, generically, to survive. In fact, for numerically small populations, one phenomenon called 'genetic drift' or 'neutral theory of evolution has been illustrated. According to these theories, in small populations, the selection does not always reward the positive solution of the organism most suitable to survive, but simply the most fortunate. (40)

It could also happen that, albeit in a vast numerical pool of individuals from a population even uniformly distributed, it may be the victim of an accident the most suitable for survival organism. In the human race, hypothetically, a possible superman could be hit by a car before procreating and spread to the offspring a mutation that is truly favorable for supremacy over other men. Sometimes more favorable mutations can occur than in other populations of the same species. Still, those geographically isolated could then be extinguished together with their superiority only for accidental local causes. Then, the fittest does not survive in such chances, as natural selection generically states.

A statistical phenomenon very similar to genetic drift is the one known as the 'Founder Effect.' When minimal populations, or part of a larger population, colonize a new environment colony may carry a specific sample that is not fully representative of the parental gene pool. Therefore, in the newly colonized area, it is possible to find individuals carrying rare or non-adaptive genes. This phenomenon occurs when the speciation processes are very intense and is defined as the 'the founder effect' or 'colonizer effect.' Not surprisingly, it is widespread on the islands.

In this regard, a well-known and significant example is the one given to us by the 'Old Order Amish' of Pennsylvania in the United States. In this group, today consisting of about 17,000 individuals, particular malformations are found with high frequency, consisting of forms of polydactyly and dwarfism, due to the action of a particular gene present in many individuals mentioned above. Since its foundation in 1700, 61 chances of this malformation have been found in this community, a number roughly equal to the total number of chances found worldwide in the same period. The cause of this phenomenon was undoubtedly the effect of the founder. The whole group was founded by only three Irish couples who moved to the United States, one of whom by chance must have been the carrier of the gene in question. This gene would later assert itself due to a series of crosses and homozygosity among the members of this community, which was, moreover, closed for religious reasons to possible marriages with members outside the community itself.

Of course, the fact that the fittest to survive does not always survive does not invalidate the explanatory value of the theory of natural selection but only its value when taken in the sense of a single guiding and predictive criterion. The fact that natural selection can survive not simply the most suitable of all, but sometimes the most suitable in fortunate circumstances brings us back to a question about the causal genesis for such a decisive event as being alive. Then the problem of the explanation of that particular type of cause, characteristic of primitive thought, reported by Pritchard (see page 8), who wrote of the existence of two different principles of causality among the primitive tribe of the Azande, reappears: on the one hand, the one aimed at understanding why a particular accident happened, linked to causes of a 'rational' and scientific type and on one other hand, the question as to why the accident happened to a specific person and not to another or, even because in some particular circumstances instead of others. For latter situations, particular and mysterious causes are invoked in primitive thought. In evolutionary terms, the question will be: why would the fly be privileged against a less annoying insect? It can be scientifically described how evolution came to the fly generation. Still, the evolutionary tree has probably had countless branches of development that are equally valid from an adaptive point of view without particular privilege for flies.

The theories of the creation or evolution from shapeless inorganic matter to living organisms are theories on the causes of phenomena or events that are not questionable. Like happens for "magical" explanations of Azandas people since it is possible to affirm everything and the opposite of everything without the practical possibility of being denied. Such explanations are, therefore, not strictly scientific theories but thought patterns. Naturally, the miraculous idea of evolution, understood as a rigid mechanism used by a God who contradicts his laws and directly determines which species should appear during history is rejected by the scientific community. Also, that hypothesis is contradicted, considering the enormous waste of lives and means deployed in the course of evolution. It appears so, is useless and incompatible with a project if targeted directly, mechanical and deterministic.

In the light of today's consolidated scientific knowledge, DNA mutations are certainly considered the engine in the evolution of living organisms and their consequent speciation. So, studying the mechanism through which mutations occur means studying the real operational foundation of evolution. However, not all the implications deriving from the modern discoveries of molecular biology have been sufficiently deepened. This deficiency requires a scientific review of creationist theories compared to Neo-Darwinist theories and the role in the natural selection of the randomness in which genetic mutations operate.

This review can be effectively justified and illustrated by examining the molecular mechanisms through which mutations occur for altering the genetic code defined in the DNA structure. Some mutations which involve considerable genetic alterations, such as in 'frameshift' (movement on the DNA chain due to insertion, and not a replacement, of a single nucleotide), involve the generation of meaningless DNA in the absolute generality of cases. While the mutation by substitution called SNP (single nucleotide polymorphism: substitution of a single nucleotide) can lead to the consequent substitution of a single amino acid to a new genetic sequence that is not always necessarily negative; these are the mutations that occur more frequently. These latter point-like variations produced in the genetic code account for various phenomena present in phenotypically similar organisms. The most obvious example is the phenomenon whereby some drugs have the expected effect on most individuals but have no effect on others. Sometimes drugs could even be harmful. Those different effects are due to some point substitutions (SNPs) in the DNA that encodes some enzymes or receptors responsible for the functioning of the active drug molecule. The branch of molecular biology called pharmacogenetics deals with these phenomena.

Of course, there are also non-lethal genetic mutations, and these modifications of the genetic sequence are often not apparent. Still, in conditions of geographical separation, they accumulate over time until they present detectable genetic, morphological, and phenotypic changes, creating differences between genetic lines and becoming gradually so deep as to define separate species. From a statistical point of view, these modifications occur at almost regular intervals. That has allowed the creation of an archeology branch, defined as genetics, which studies the emergence and development of differentiation and separations between the various populations in their evolutionary history. The greater the number of mutations different from each other in comparing two groups of similar organisms, the further back in time the genetic separation from a common ancestor took place. By studying DNA extracted from ancient finds, population genetics scholars have managed to reconstruct a sort of genealogy between the various human populations that have appeared on earth, confirming or disagreeing with the reconstructions based only on the various archaeological finds and fossils. Genetic archeology, therefore, rests its predictive abilities on variations in genetic populations related to statistical regularities in the frequency of the onset of mutations. Of course, this is a purely quantitative predictive capacity; it foresees how many mutations have arisen. It does not appear qualitative; that is, it does not predict the peculiar emerging organic characteristics, so it is not sufficient to give a definite account of the onset of a certain diverse species. However, it is possible to predict mutations that are certainly lethal therefore impossible to transmit to progeny and that block many evolutionary paths.

For example, despite being inscribed in the list of mathematical probabilities of mutation, an insect cannot generate offspring of the same species as large as a man since insects possess only an external skeleton. That, for terrestrial gravity, proves to be fragile, unlike other animals that possess an inner skeleton capable of withstanding large size and weight. Furthermore, an animal cannot be too large without a respiratory system of some kind because oxygen could not reach the body by simple gas permeability without a respiratory system. Trees cannot grow beyond certain heights because the sap coming from the roots cannot reach beyond the limits dictated by the force of gravity, etc. On the other hand, if we go outside the laws of nature, we could glimpse an evolutionary path of man toward fantastic superheroes with powers such as flying by levitation, emitting fire, crossing walls, making themselves invisible. Not possible things that represent a creative funnel, a filter to evolution making it partly predictable, since it can only move on possible paths within the laws of nature that already, as such, allow for wonders.

The idea of predictable evolutionary paths is supported by the phylogeny of the central nervous system in the animal world. The origin of the nervous system is, at first glance, observed in the most primitive coelenterates animals, where the sensory epithelial cells become differentiated and in which excitability is enhanced. These cells have a contractile offshoot in the deepest part and are called myoepithelial cells. They detect stimuli and are capable of processing a motor response and carrying it out directly. During the phylogeny, a separation is created between the sensory and contractile elements, which are located more deeply by connecting with an extension in which a new formation called motor neuron is interposed. This primitive cell is the first nerve cell. The subsequent structural evolutions can be schematically seen in fig. 1 and are familiar to the animal world in general. A functional path accompanies the structural evolutionary path. From the first chemotactic tropisms, we pass to the stipulation of

The evolution engine

S-R systems (stimulus reaction) to develop developed assessment and cognitive faculties. A morphological evolution then leads to cephalization of the CNS, positioning the primary sensory organs such as sight, smell, and hearing next to the brain. So, in the direction of organism's locomotion for reasons of functionalities related to the rapid response of the animal organism to the stimuli he finds in its path.

Also noteworthy are the phenomena of biological convergence, which involve the channeling of different organisms and species into similar forms when in similar environmental conditions. This phenomenon occurs, for example, in desert areas where plants take on similar shapes in the leaves and structure, in general, to reduce the surface/volume ratio of the organism, reducing transpiration and water loss. It can therefore be seen that, although all mutations are linked to the phenomena of randomness, they do not, for this reason, lead to necessarily unexpected developments. The same happens for ideas presented in some theories that, due to a similarity of chemical properties, also hypothesize that life can emerge from silicon such as that based on carbon, but that is impossible. In that, due to chemical limitations, randomness would fail in it. Considering these considerations, a higher intelligence, as a cause that remains in effect, could undoubtedly foresee and define not by miracles but by the intrinsic natural laws, all the variability implicit in the genetic code, and also a final way between the various genetic lines up to the genetic line of Homo sapiens among others. Through its own free and unpredictable choice, the latter could then upset the future evolutionary path that was previously in some way predictable until its appearance on the scene of nature.

173

The evolution engine

The phenomenon of mutations in living organisms also raises considerations fundamentally different from those concerning molecular diagnostics or the search for our genetic roots. One of the most incredible mechanisms in genomic mutations is represented by the branch in molecular biology called transposomics, a discipline born in the late forties of the last century, thanks to Barbara McClintock. Transposomics studies a genome coding that allows a particular operating method by creating a molecular mechanism designed to generate alterations in its genomic code. This phenomenon undermines a cornerstone of previous beliefs that saw the defence of genetic stability and its copying processes as one of the most important mechanisms for species conservation. That is because the scientific community thought that the genome was immutable, except for random copy errors or errors induced by factors external to the genome. Today, surprisingly, we learn that this is not the case. Instead, the genetic code also works to modify itself: transposons can move from one region of the genome to another to insert themselves in certain areas, thus modifying some DNA sequences. The discovery of transposons, also known as jumping genes, has revolutionized genetics by radically changing our understanding of cell differentiation through mutations. Far from being a helpless mechanism linked to weird transcription errors of the genetic code, the transposomal molecular mechanism is structural in the genome with the function of causing important mutations such as deletions, inversions, and translocations.

Transposons are found in both prokaryotes and eukaryotes: they are located on the same chromosome coding for their transposition. In eukaryotes, the question is similar but more complex since they can move to different positions in the same chromosome and different chromosomes. Today, we speak more broadly of "transposable genetic elements" since different classes of genetic elements can vary the genomic sequence. This definition allows embracing the whole complex of living beings since all have transposable elements capable of producing mutations.

The transposomics mechanisms capable of producing genomic rearrangements are astonishing because they demonstrate how, in this way, nature has produced mutations capable of inducing other mutations. With enormous surprise, biology has discovered the onset during the evolution of a molecular mechanism suitable for altering its codes to arrive more quickly to produce other mutations. Even more surprisingly, natural selection preserved that mechanism. The positivity of a mutation for different organisms is naturally expressed as a better adaptation to the different and changing environments that living organisms face. In his cited book, Monod observes how faithfulness in reproducing the genetic code and its transmission to the offspring is genuinely astonishing and necessary. Of course, this is entirely correct, but the discoveries by modern molecular biology of the enormous importance of the mechanisms responsible for mutations are even more impressive. These show that mutations and change have greater importance in the structure of our genetic code than copy fidelity. In fact, contrary to what one might suppose, shares of the genome used in mutations mechanisms are much higher than the shares of the genetic code responsible for coding the molecules necessary to the organism for its functioning.

The evolutionary process of producing mutations mechanisms relevance can be effectively represented by the percentages of operational DNA, useful for phenotype, present in the human genome. It has been found that about 45% of the genome consists of transposable DNA elements. In contrast, the coding DNA, which produces the enzymatic and structural proteins necessary for the human body, represents only 25,000 genes, 1.5% of all DNA. Almost 53% of the total genomic consists of non-coding DNA such as introns and structural DNA, repeated sequences, segmental duplications that are sometimes referred to as 'Junk DNA.' There are, furthermore, nucleotide traits that could have a functional significance acting as regulators that, while not directly intervening in the coding responsible for protein synthesis, would act by activating or deactivating adjacent DNA segments. Recently, in a study conducted by an international team led by researchers from the School of Medicine of San Diego, University of California, a positive and statistically significant correlation was discovered between the alterations of this so-called junk DNA and the presence of autism. These variants do not occur as new mutations in children with autism but are inherited from their male parents [41]. These studies show how non-coding DNA could intervene to regulate complex behaviors in organisms, not otherwise comprehensible on the sole basis of the availability of directly coding DNA.

While many mutations such as SNPs can be explained as simple copy errors during DNA replication processes, a question about this distribution of DNA arises spontaneously. Why is there such an imposing presence of information in the genetic code only to produce mutations that aren't simple copying errors? Why are "deviations on what to copy" induced with such massive use of genetic resources?

The evolution engine

Many biologists usually give one stereotyped answer when someone, amazed, asks why exists such perfection and usefulness of biological processes. It consists of affirming that chance in the evolution of living organisms, even among many failures, produces mechanisms and biological structures that are ever more perfect and suitable for the survival of the individual and the species, without necessarily having a design by a creator. An answer of this kind to explain the appearance of a mutation in the genetic code designed to produce further mutations appears challenging to propose as exhaustive. It doesn't seem acceptable an explanation where perfect randomness, through mutations, has selected a mechanism for which it produces mutations to alter the species' future evolution of the species? A sort of genetic system that we can call a meta-mutation process. Everything is possible, so an answer of stereotyped neo-Darwinism to this question is provided by invoking that the increase in possible mutations can accelerate the pace of natural evolution towards ever more perfect living creatures and with a consequent adaptive advantage. However, an objection to this explanation is that organism mutations always lead to immediate adaptive improvement promptly reflected as an advantage to the offspring. A meta-mutation, instead, does not find confirmation in the immediate offspring but should find it in a generally subsequent evolutionary path, adapting in the future more quickly to the environment it will find. Indeed, the acceleration in mutational processes should produce in the future more prompt responses to any environmental changes and, even lacking these, new advantages for the expansion of the species. However, a genuinely random mutation does not have any projects. Therefore, it does not have a plan to carry out that ultimately foresees a successful future through the acceleration of natural evolution. An investment that should then be repaid over time, as happens for the capital investment in an industrial enterprise. This foreseeing would only be possible for an abstract thought process that imagines future benefits.

Furthermore, If it is devoid of immediate meaning and if, due to a sort of genetic inertia, such a mutation were preserved by nature, it would not be preserved for so long as to extend from bacteria to mammals. The good for the species should be an adaptive advantage immediately from the moment of the mutation. Otherwise, it would be an advantage towards something in an unimaginable future, an unthinkable fact for an evolutionary process that should be unconscious, as represented by neo-Darwinism, and an actual objective representation of nature.

So, the genetic code fidelity stabilizes the species for longer or shorter periods, thus replicating the Trojan horse, represented by the transposomic genetic mechanism, despite its uselessness for the immediate environmental conditions. The transposomic mechanism destined to produce mutations can be defined as a meta-mutation since it occurs not as a normal adaptive process but to produce other future mutations. That can be seen as an advantage to producing and accelerating other meta-mutations, increasing the probabilities of good variations since the valuable and vital mutations for the organism are increasingly difficult to obtain considering the incredible complexity necessary for the metabolic and functional balances achieved in living organisms. However, this utility can only be conceived by an observer external to the evolutionary process as a whole and is not conceivable by a blind and objective evolutionary process based on chance and the exact process of copying. On the other hand, permanently useless genetic coding is eliminated even if not immediately harmful. Only an intelligent project can exchange a present sacrifice for a future good to be defined and evaluated.

The elimination of the chance from evolution

The elimination of the chance from evolution

Man's work is evaluated to eliminate random and unwanted imperfections from the genetic code of living beings to improve their chances of survival and the ability to conform oneself to dominant or otherwise desired aesthetic models.

Man has always tried to eliminate in the surrounding environment the circumstances and living organisms that appeared unfavorable to him to create or maximize those favorable for his survival and well-being. From a practical point of view, man's intellectual and cognitive superiority over animals is represented not so much by his ability, often also common to higher animals, to use tools but by his ability to build them. This capacity has allowed man to modify the surrounding environments to adapt them to his own needs rather than escape from environments that are little or not entirely suitable for him.

Today genetic engineering intervenes to modify the genome of different organisms to adapt their genetic code to the different needs of man, especially in medicine and agriculture. In recent times, the manipulation of the different genetic sequences has taken place in the medical field to correct the human genetic code to eliminate any defects in the nucleotide sequences that result in diseases for the carriers of these anomalies. The different applications for modifying the genome in other natural species concern the production of some organic molecules with pharmacological activity, which are difficult to obtain by standard chemical methods. Some drugs obtainable from genetic manipulation are Erythropoietin which stimulates the production of red blood cells in the bone marrow, growth hormone (Growth hormone: GH), and, more recently, also atrial peptides useful for the treatment of hypertension. Arterial such as TPA (Tissue Plasminogen Activator) is a human protein typically produced in the body in small quantities and is particularly effective in inducing the breakdown of blood clots. In the agricultural field, the manipulation is carried out to form GMO plants (genetically modified organisms), which acquire characteristics that allow greater productivity and more excellent resistance to parasites and pathogenic bacteria. Known examples are represented by Golden rice, a variety engineered through the insertion of a segment of DNA that gives it greater nutritional capacity and a high content of beta-carotene, which is then transformed into vitamin A in the human body. Another case is represented by BT corn, in which the insertion of foreign DNA into its genome gives it the property of producing a toxic substance for most harmful insects.

The path to unveiling stretches of DNA that encode functional proteins is now relatively easy because the amino acid sequences of the proteins produced and therefore the nucleotide sequence are known, while it is not possible, on the contrary, to create "ex Novo" nucleotide sequences that encode complex proteins with of the desired and pre-ordered characteristics if not already pre-existing: in other words, we can say that we can read the genetic code but not to write it authentically, that is, not with complete meaningful sentences. At least today, there is no possibility of designing a complex genetic code capable of expressing new functional proteins starting from single nucleotides. It is possible only to copy already known and pre-existing sequences in other organisms, as happens in some cases of genetic engineering. These sophisticated techniques use only so-called copy/paste techniques for nucleotide sequences with known functionality. Because of this inability, in terms of genetic planning, it appears extremely dangerous to blindly modify genomes, such as viral or bacterial ones, due to the risks associated with the real possibility of creating super-pathogenic viruses or bacteria in terms of replication speed or infectivity. And maybe even super resistant to the drugs available.

The manipulations performed mainly in reproductive cells have been carried out at a sub-cellular and non-molecular level. This type of operation is mainly represented by replacing the original nucleus of the oocyte with another nucleus suitably treated and derived from a somatic cell. The operation gives rise to the development of a clone of the organism to which the somatic cell belongs, from which the foreign nucleus that has replaced the host egg original nucleus is obtained. The procedure leads to viable clones but is often affected by some defects not yet identified. The first chance, which caused a stir in the media, concerns the cloning of the sheep called Dolly, which, despite being an identical copy of the original 'mother' sheep, presented various anomalies diseases for which she died prematurely.

Molecular biology has revolutionized the life sciences with applications that have often not had negative repercussions on ethics or caused an evident social impact because of used in the branches that deal with molecular diagnostics, forensic medicine, population genetics, and pharmacology. There were several considerations and perplexities raised by civil society when some applications of genetic engineering, typically used in agriculture to produce genetically modified organisms, were transferred to the possible modification of the human genome (Human genetic editing).

Considerable progress has been made in modifying the genetic code when, recently, particular endonucleases (enzymes that cut the DNA chains on intermediate sections of the nucleotide chain) were discovered and called CRISPScas9 'Clustered Regularly Interspaced Short Palindromic Repeats'. CRISPRs are part of the immune system of microorganisms discovered twelve years ago in the bacterial flora of yogurt. These complex molecules have allowed careful editing of the genome. They are composed of different sub-units that can cut DNA, no longer only in predetermined, non-modifiable areas, as happens with the restriction enzymes used for these operations, which cut only in fixed points of the nucleotide sequence. CRISPRs can cut the DNA in eukaryotes and humans at points decided by the biologist to eliminate traits of the gene that are considered abnormal or cut out traits for subsequent uses. In the chance of genetic diseases, 'wrong' sequences can therefore be replaced by biologists with "right" sequences to create a normally functioning gene capable of producing the correct proteins or enzymes. These molecular scissors are the tool of a nascent gene therapy technique that should come to treat diseases such as sickle cell anaemia, Huntington's disease, cystic fibrosis. This scissor can

also introduce genes that offer protection to predisposing family-based diseases such as heart attack, Alzheimer's disease, and other genetic predispositions towards some cancers. Operators could perform the malfunctioning gene replacement in the embryonic phase on the male or female gametes. So that cells replicate in the organism growth phase, all containing the correct gene that can be transmitted to the progeny. The selected genetic traits could also be inserted into somatic cells of the adult organism, therefore not transmitted, with a curative effect limited to the single individual's existence.

Of course, gene therapy also has an aspect of gene editing, which, theoretically, can serve from a clinical point of view and produce effects that are considered advantageous, such as intelligence and other special abilities. Another goal could be to give the offspring other desired phenotypic effects such as skin, hair, eye color, height, or others. Considering only some of the desirable characteristics, thus opening the way to a sort of eugenics that is functional and aesthetic.

State of the art regarding this branch of molecular genetics indicates that the introduction of DNA manipulative techniques has already paved the way toward scenarios that are, to say the least, disturbing for the ethical and human health implications. In fact, after numerous experiments on animal organisms, researchers have subsequently moved on to manipulations of the human genome. Such as the one conducted by Junjiu Huang [42] performed on embryos in the zygote stage before cell division at the time of the pronuclei formation, which must be two, representing the genetic heritages of the parents. These experiments, however, were conducted on zygotes that presented with three pronuclei instead of two and, therefore, doomed sterile. Later on, there was the shock announcement of a practical application conducted on viable embryos. On Monday, 26 November 2018, during the International Summit on Human Genome Editing, underway in Hong Kong, the Chinese scientist He Jiankui announced that he had conducted an editing test on human embryos that resulted in the birth of twins with modified DNA. The initial bewilderment had led many in the scientific community to doubt what had happened. However, after attending He's presentation and listening to answers to dozens of questions formulated by the scientists in the room, there was then little doubt the experiment had been conducted. The Chinese judiciary court condemned these practices, so the researcher was sentenced in January 2020 to three years in prison for "illegal medical practices," with two of his collaborators sentenced to fines. There were also notable protests in the scientific and political world to the fear that some biotechnology companies were already thinking about monetizing this genetic modification technique.

In March 2015, Edward Lanphier and four other researchers asked for a moratorium on the biomedical use of new techniques, which are still immature and risky. A point of view that appears to represent well the current thought of scientists engaged in this research field who declared: "In our opinion, the genome editing in human embryos using current technologies could have unpredictable effects on future generations. This risk makes it dangerous and ethically unacceptable. Such research could be exploited for non-therapeutic modifications. We are concerned that a public outcry about such an ethical violation could hinder a promising area of therapeutic development, namely, making genetic changes that cannot be inherited. At this early stage, scientists should agree not to modify the DNA of human reproductive cells. If a truly compelling chance arises for the therapeutic benefit of germline modification, then we would encourage open discussion on the appropriate course of action. " [43]

Due to their relative simplicity and speed of reproduction, bacteria are one of the organisms of choice for genome modifications. It has recently been possible by researchers led by Julius Fredens to synthesize an entire Escherichia Coli genome of which the entire nucleotide sequence is known. The synthesis of a genome of the E. Coli bacterium has been described with the use of 61 codons - triplets of nucleotides - producing the same type of protein as happening by the use of 64 codons as occurs in nature to encode the 20 amino acids but with a certain redundancy of available codes. Therefore, it has been shown that the number of codons used to encode canonical amino acids can be reduced, apparently without risk. The method described allowed the creation of a variant of Escherichia coli with a synthetic genome produced through a high-fidelity total synthesis of the nucleotide sequence. The synthetic genome thus obtained was composed of 18,214 codons which allowed to create of an organism with a genome of 61 codons; this organism uses 59 codons to encode the 20 amino acids (two codons are of the regulatory type) and thus allows the deletion of a previously essential transfer RNA. The researchers then demonstrated that this E. coli strain is viable and behaves normally like its natural counterparts even though it has lower replication rates than the original strain. [44] According to biologists, this research opens the way to the design of synthetic bacteria capable of producing substances that cannot be obtained by biosynthesis starting from microorganisms in the natural state.

Of course, the DNA codes in this chance encode a sequence of amino acids equal to the original; however, it is an inexpensive copy that exploits the redundancy of the amino acid codes. This reproduction still concerns a copy, not an original microorganism creation. As is known, we can only read and copy but not write codes of DNA nucleotide sequences of organisms for new genuine invention, which

The elimination of the chance from evolution

can then lead to the production, not random but useless, of proteins. Despite what can be defined as a considerable technical success, also in this chance, as for others, it seems entirely necessary to exercise all caution today in not wanting to write for invention, but only to copy the existing without deviations from the original with blind changes.

The homination

The particularity in the evolution that led to the Homo sapiens species, unlike other species, is identified with its genus despite its high rate of mutation, which would lead to suppose the existence of many existing contemporary species. Perhaps they disappeared with the emergence of the consciousness of an afterlife in time and space, which foresees dangers for one's survival far beyond what is immediate and local.

The latest product of the homination process is represented by Homo sapiens, which with its baggage of mutations, has been led to differing from other Homo species and has occupied all possible biological niches. Homo Neanderthalensis and Denisovans, the last species Homo sapiens have coexisted with, have left genetic traces on their genome, crossing with limited but appreciable percentages. Those species could also be considered as breeds due to limited interfertility. There may also have been crossings with other archaic populations, especially with different genetic populations outside the European continent, currently less studied by population genetics.

The genus Homo is part of the order of primates composed of 15 families, divided into numerous subfamilies, genera, and species. However, Homo sapiens is a species that identifies with its Genus despite having had numerous speciation in its evolutionary process from antecedent species. Therefore, we can say Genus or human species, thus indicating the same genetic population. In the family Hominidae, we find the subfamily Ponginae with several species still living. The subfamily Homininae is divided into the Gorillini tribe with several living species, and the Hominini tribe, which is divided into the genus

The homination

Pan with two species, and finally the Genus Homo with only one species. While numerous species and genera have survived, cohabiting, with the other primates, for the genus 'Homo Sapiens,' we are witnessing the disappearance of all the species and races that preceded it; therefore, all ancestors and close relatives are absent. In particular, all the ascendant relatives of man after the Chimpanzee, which has been separated from man by millions of years of evolution, have become 'extinct.' So, about possible close relatives, no one currently coexists with modern man. That perhaps suggests the idea of finding ourselves in the presence of a warlike and intolerant species toward other species similar to him, which could threaten his ecological niche. The disappearance of human close genetic relatives probably also lies in the fact that the environmental niche became exploitable to humans could extend throughout the planet. In a particular way, man could extend the boundaries of his biosphere over time, too, since he can see present dangers and imagine future dangers.

Tracing the dates of man's evolutionary path is difficult due to the wide margins of oscillation granted by the technique that uses for the density of mutations offered by genetic analysis or, even if with more precision, in the dating by the decay of the carbon14. The dating concerning the differentiation between species and genera is therefore completely approximate. It is also influenced by the always possible discovery of new fossil finds, making us reconsider acquisitions and dates based on other previous findings. Different authors sometimes report even different data on the same evolutionary junction. Therefore, each historical description can't be complete and precise in fine detail.

The homination

In the previous period, between 8-10 Ma (Millions of years), in the African continent, the evolutionary line of the Chimpanzee began to diverge from that which led to the current Homo Sapiens. From the archaeological and genetic research carried out, it is possible to trace a list of the species passed through a temporal sequence. The detachment of the hominids from the same genetic trunk shared with the Chimpanzee descends at about 7 (Ma).

The detachment from the ancestor occurred with the differentiation of the Orrorin tugenensis species, coming from Kenya, which showed the possibility of walking in an upright position, with highly pronounced primitive characteristics, but which probably led to the evolutionary line of the australopithecines. Approximately 5.5-6.0Ma lived the Ardipithecus kadabba, and 4.4Ma lived the Ardipithecus ramidus, a mixture of hominid and monkey; 3.25 But Australopithecus afarensis lived, of which pretty representative fragments of 'Lucy' and other artifacts known as the 'first family' were found. Then the Australopithecus africanus 3-2Ma appeared almost at the same time. Then Paranthropus 2, 75Ma and Australopithecus Garhi 2.5Ma.

The genus Homo begins its evolutionary path around the period dating back to 2.5Ma. The Homo habilis considered the first known species of the genus Homo. In chronological order, and always in broad terms, Homo rudolfensis 1.9Ma evolves the Homo erectus 1.75 Ma, Homo ergaster 1.75Ma, Homo Naledi 1,5 considered, due to anatomical differences an intermediate between the genus Australopithecus and Homo. Then appeared the Homo antecessor 1.25Ma; Homo heidelbergensis 0.6Ma, which will become extinct after 400,000 years;

The homination

Homo Neanderthal and Homo Denisovans 0.6Ma, which will become extinct about 40,000 years before the present time and Homo sapiens 0.2Ma. The Homo sapiens idaltu is also reported about 160,000 years ago, considered an archaic subspecies now extinct. The Homo "Red Deer Cave people" 11,000-14,000 years old, whose fossils were found in China in 2012 and are of uncertain classification. In the same period lived Homo floresiensis, a tiny human species, due to insular dwarfism, extinct about 12,000 years ago found in 2003.

By narrowing the observation period to see in more detail, it is observed that the closest relatives of modern man in time were Homo Neardenthalensis, Homo Denisovans, and Homo Floresiensis. The latter, whose fossil remains were recently discovered in 2003, was also called "hobbits" since, at that time, a film "The Lord of the Rings" was released which spoke of men who were utterly equal in cognitive abilities to the other inhabitants of the earth but very small in size. Floresiensis did not exceed one meter in height, even though they possessed a remarkable genetic similarity with Homo sapiens. In Europe, for Homo sapiens, the gene component due to Neanderthal has been detected from 1 to 4%, while this gene component is not present in the original African populations. There is also a genetic overlap between the Denisovan genome and some current Asians, particularly accentuated with a group of Pacific islanders living in Papua New Guinea known as Melanesians. It seems that Denisovans contributed between 3 and 5 percent of their genetic material to the genomes of Melanesians. It is believed that the Denisovans, who lived in eastern Eurasia, probably interbred with the ancestors of the Melanesians when they were about to cross the ocean and reach Papua New Guinea about 45,000 years ago.

The homination

According to the most accredited research, the Neanderthals, Denisovans, and today's Sapiens all descend from the oldest Homo Heidelbergensis with a genetic divergence that began between 500,000 and 400,000 years ago. Some genetic research suggests that an ancestral group of H. Heidelbergensis left Africa and then differentiated shortly after that with a branch that ventured northwest into Western Asia and Europe with a subsequent mutation to Neanderthals. At the same time, another branch moved east, becoming Homo Denisovans. Homo Heidelbergensis, about 130,000 years ago, had become a further genetic branch in Africa, namely Homo sapiens Cro-magnon which began its departure from Africa about 60,000 years ago.

Those who see an intelligent design inscribed in the evolutionary process of man put some observations supporting their thesis. The first consideration concerns the species rate extinction after the chimpanzee up to the sapiens species because of differences in average extinction rates observed between one species and another. It is for the man about 300,000 years, while the extinction rate of the other species within the class of mammals is estimated between 500,000 and 2 million years. That is a phenomenon in disagreement with natural competition for which species with much higher cognitive faculties, compared to other animal species, should have many more chances of survival and permanence. It seems it should happen for humankind thanks to his adaptability. The second observation, linked to the first, concerns the characteristic of presenting in the evolutionary line that leads to Homo sapiens a genome with a higher mutations rate than other evolutionary lines, making the evolutionary process faster. This observation has led to the onset of many human species, while the genetic lines of species close to our evolutionary line are far more stable. It has been seen that the Homo sapiens species is the only living species left of its genus, leaving behind at least 20 different species throughout 6 million years, a number that is undoubtedly destined to increase due to possible other fossil finds.

The Chimpanzee (Pan troglodytes), which shares the same ancestor as man, has remained stable for an extended period of 4.5-5.0 million years. 'Only' 1.8 million years has separated from the same genetic branch, the Pan bonobo, a species of smaller size but still with primitive characteristics. The Gorilla genus was stable for about 8Ma; Orango remained stable for about 6Ma. By comparison, these data demonstrate the relative great mutation capacity of the genus Homo. For the theories that support a purpose in nature, the disappearance of species, even brilliant compared to other animal species, reflects an initial plan by God to arrive at definitive Homo sapiens, a creature with the faculty of reflecting on its Creator. It is the only one fully capable of not simply stopping at the immediately sensitive or otherwise experienceable datum.

Other points of view can interpret these characteristics of the evolutionary line of man as winning adaptive mechanisms that operate in favor of species that evolve very rapidly. That is thanks to a greater and formidable mutations capacity, not present to such a high degree in other animals of the same genetic family. A genetic mechanism, such as that represented by transposomics, would give a genome a more rapid mutation process which could improve the chances of survival of a particular evolutionary line by offering a more comprehensive range of possibilities to natural selection in the moment of rapid environmental changes. An example of such epochal changes is represented by the alternation of temperatures on the earth. Through modern methods carried out with the coring of the ice and measurement of the isotopes present between the ice layers, it has been calculated that the earth has gone through seven glaciations. From the end of the lower Pleistocene (1,2 million years) until the last glaciation ended about 12,000 years ago, periods interspersed with interglacial periods in which the temperature rises considerably. The last interglacial period is the one we are going through, which obviously will not end soon. The distances between one epoch and another are measured in hundreds of thousands of years. Organisms capable of mutating faster can offer more significant possibilities to produce a relatively greater variety of organisms, among which those most suited to the ever-new conditions will survive. Of course, an impersonal evolutionary process can select the species best suited to the environmental conditions of the moment regardless of how many losses can be suffered through selection for disappearances of individuals and populations.

Perhaps a slightly different consideration can be made regarding Homo sapiens and the disappearance of previous generations: the evolutionary picture could be a particular chance of natural selection in which the strongest has eliminated 'the other' since, beyond a certain intellectual degree, man can conceive the other species as a deferred danger, even if not immediately present as a threat. The selection was thus also able to operate in an almost 'personal' way, eliminating the weakest, by cognitive perspective, even when apparently brother's species. On the other hand, the bloody wars throughout human history are perhaps not inferior in cruelty to those indifferent reported through prehistory.

The wonder of life

THE WONDER OF LIFE

In human wonder, the feeling is often evoked in contemplating living beings' beauty and functional perfection. According to the concept of natural selection, wonder is not justified as it would turn out to be a particular state of mind and simply an irrational emotional residue.

The man of religion looks at nature as a perennial theophany, the manifestation of the word of God that is expressed in living nature. He sees beauty as the echo of light of the intelligible into the sensible. Hence, the wonder and praise to the Lord. The atheist is equally astonished, but he doesn't have to thank anyone since the beautifulness should be attributed to a state of amazement and admiration of a technical nature without other emotional involvement. It is something that only for insufficient scientific progress of the moment is impossible to explain in all detail.

197

The wonder of life

Speaking of wonder, we cannot avoid quoting Aristotle at the beginning of his first book on metaphysics: "*All men are naturally inclined to knowledge, the joy they feel for sensations is an evident sign, since these, even if they are put aside the usefulness that derives from them, they are loved for themselves, and more than all the others the one exercised through the eyes is loved.* " It is evident here that the senses are the bearers of our knowledge. We take pleasure in acquiring as happens, for example, for Archimedes' 'eureka!' when he catches the understanding of the law contained in the principle that bears his name. Scientific knowledge of the mechanisms underlying the functioning of living organisms may, perhaps, eliminate this feeling through the force of rationality. Life can then appear as the result of a mechanical process that is simply explainable, as the existing physical laws regulate it; therefore, no ineffable wonder seems to be justified.

Modern biology has variously described life in terms of the mechanisms and laws underlying the processes that operate in living structures. The description of these processes remains linked not to 'why' but to 'how' living organisms' function once they have reached being by emerging from the inorganic material world. However, the emergence of living organisms from the inorganic world appears to be one of those problems that human reason cannot help but tackle by seeking a scientific explanation that, as such, goes beyond description. Therefore, the man of science or religion seeks to eliminate the doubt placed at the crossroads between chance or the need for a divine plan as possible alternative causes for the origin of life.

It has already been widely written about the problem linked to the probability of emergence by pure chance of living structures starting from the inorganic world. But, the problems posed by the evolution of living beings alone are entirely understandable and certainly easier to solve than the possible mechanisms operating for the first emergence of life. The widely accepted theory of the neo-Darwinian paradigm offers an adequate scientific answer to the evolution of living beings. It is based on chance, which represents the mechanism and the objective interpretative criterion for evaluating what appears inexplicable and incredible at first sight. The neo-Darwinian conception of evolution has erased all wonders by making every feature, progress, or eventuality that can be found for every living form, right from the simplest organisms, fall within the scope of a clear and ordinary technical explanation.

Those who deeply know the perfection and incredible complexity of structures, metabolic balances, and instinctual behaviors in higher organisms wonder how these attractive characteristics are possible. While understanding the mechanism of natural selection, some biologists and even non-specialists in life sciences also dwell on a question that arises upstream of this mechanism: that is, how all this is possible without a mind that directs and produces such perfections? A kind of mantra is always proposed to these questions for every possible marvelous emerging characteristic of a living being. It reads: "this characteristic represents an adaptive advantage that natural selection has conferred on the organism, and is then transmitted to the entire species in subsequent reproductive crossings." The same answer is given for all the essential functions that the organism must show.

Nature expresses all possible combinations for every form of life through the onset of random mutations. After Darwin, it became clear that the mutated living forms, produced in large numbers that eventually survive, compared to others that disappear, propagate, and occupy their particular environmental niche because they are the most suitable for the environmental conditions of the moment. From this, of course, no wonder: it is simply a mechanism of natural selection that is understood with a logic of common sense by looking at the number of life forms that die. It is stated by many biologists that if there were a guided and intelligent selection, there would not be all the useless hecatomb represented by the organisms that, not suitable, die.

The minimum characteristics that define the living and that we have seen represented by Gerard Karp on pages 61/62 do not arouse any surprise for them to occur in nature. If these characteristics did not exist, we could not have a living organism, so no wonder above all for the essentials possessed even by the simplest of living organisms. In the list of characteristics mentioned by Karp, reproductive capacity does not appear, which ultimately does not necessarily have to belong to a living being in the strict sense. However, all living beings must possess this property to not be a simple episode within the framework of nature and without an evolutionary outlet.

Konrad Lorenz exemplifies the sometimes very sophisticated behaviors present in different animal species in his writing "the other side of the mirror," a title that stands as a metaphor for the physiological apparatus of the living. He notes that in geese, the attachment to the parent is a genetic behavioral seal called 'imprinting,' that is, a stereotyped and innate pattern of behavior. Following a predetermined pattern, the newborn offspring recognizes an animal as its parent, not according to the species. Still, according to the first visualization concerning an animated being: the first living being they see is taken as a parent to follow, even if it were a man. The genetic message is clear; by imprinting, the newborn recognizes its primary source of food and protection, thus making possible or even increasing its chances of survival. Even in this chance, no wonder for such complex behavior. Since it is simply a mechanism driven by a winning flexible genetic scheme over too specific a recognition, this non-specificity allows them to be protected by adoption. The newborn will have a greater chance of survival than other species lacking this recognition mechanism with too much orientation directive regarding biological parenting: this certainly does not involve relationships of filial affections, for atheists considered sublimated, present in the human family.

According to some theories, behaviors are also possibly inscribed in the genetic code in the form of so-called junk DNA, representing approximately 53% of the total genome DNA (consisting of non-coding DNA such as introns, structural DNA, repeated sequences, and segmental duplications). On the other hand, junk genetic sequences can find functional significance by acting as regulators, not directly intervening in the coding responsible for protein synthesis by activating or deactivating adjacent DNA segments. Some studies show how non-coding DNA could intervene to regulate complex behaviors in organisms, which is not otherwise understandable on the sole basis of the availability of DNA only directly encoding proteins. As with all genetically transmitted traits, it could be possible that complex behaviors code in junk DNA, entailing a very new adaptive advantage. Ethology, therefore, would not stray at all from the neo-Darwinian paradigm. So difficult to explain behaviors would result from an advantageous genetic codification, subject to natural selection to permit the onset of phenotypic characters that would contribute to the affirmation and speciation of animal populations. Anthropologists can extend the same interpretative criterion of behaviors to all the manifestations of human feelings. For such a theory, emotions and feelings would be nothing more than genetically encoded and modifiable instincts in the course of natural evolution.

Under strictly biological ethics, for a living organism, the good is what makes it live best, makes it develop, and effectively multiplies the species' population to which it belongs. A direct logical consequence of this ethics is that life originates from chance without questioning the moral value. According to this idea, all the behaviors of superior animals, even if defined in some way altruistic, are nothing more than behaviors dictated by the genetic code for the benefit of oneself as a single organism and indirectly also for the benefit of one's species.

The fundamental survival instinct requires the individual organism and its species to protect itself, assert itself and reproduce as quickly and effectively as possible. In this perspective, no behavior is genuinely altruistic. Even parents' love towards their children in higher animals does not arouse wonder since it is nothing more than a feeling dictated by the genetic code, which allows the offspring to survive better or better reproduce. In the Darwinian mechanism, even the best sentiment, such as love, falls within the typical species survival mechanism, dictated by specific genetic sequences. For this, aspect too, no wonder because is there is no providential plan that justifies the need for a creator God. Life does not seem to require a creation; everything has a scientific explanation linked to the domain of the DNA that wants to reproduce.

The wonder of life

A vivid example is brought in in 1976 by Richard Dawkins. He published the book 'The selfish gene,' from which we read: "The Darwinian' survival of the fittest 'is a particular chance of a more general law of survival of what is stable. Stable things populate the universe, and a stable thing is a collection of atoms that is permanent or common enough to deserve a name"[45]. Dawkins affirms the power of DNA, and the gene is a stable and legislative thing compared to the individual organism. So, paraphrasing the old question of whether the chicken or the egg was born first affirms the concept that the hen (or man) is nothing but the egg's tool to produce another egg. The single living organism is only a temporary passage.

From a strictly materialistic biological perspective, life does not show any direct creative intervention. We can explain every biological manifestation through the need to positively select every physical and mental characteristic capable of increasing the chances of survival and reproducing itself. Every aspect worthy of wonder is justified by adaptive necessity in the struggle for the survival of individuals and species. In the most advanced animals, the individual is endowed with a formidable instinct to survive alongside an equally formidable sexual instinct aimed at reproduction for the multiplication of the species. The more significant reproductive differential between one species and another, together with the length of the life of the single organism, is the winning weapon to increase the number of members of the species and represents a powerful tool in the struggle for dominance in one's ecological niche. The species that reproduce little must compensate for this evolutionary disadvantage with two different mechanisms operating together or disjoint: increasing parental care towards new born organisms up to their complete autonomy in dealing with the surrounding environment and extending individual life duration.

Many species do not require parental care: fish, amphibians, and reptiles lay a vast number of eggs which are in most chances left to fend for themselves. The surrounding fauna eats many eggs. The tiny creatures that could already move and search for food independently came from the remaining. Without the protection of their parents, they are easy prey for other animals, but a large number of eggs laid guarantees the continuity of the species. Parental care, conversely, is mainly present among birds and mammals: the former lay a few eggs for each brood, and the latter can give birth to a few children at a time. In most chances, the newborns of birds and mammals cannot move and get food on their own, and it is necessary that the parents provide them with food, defend them from any predators, and teach them to fly or walk. In addition to this, they also transmit many other behaviors to them, either by genetics or by imitative education, to guarantee them better chances of survival. Parental care, in most cases, is provided by the mother, but chances are not uncommon where both parents are involved in raising the children. For the human species, the number of offspring is small, and thus the duration and attention in parental care is very long, even when compared with other primates. In any case, the general rule is that the fewer the offspring, the greater the parental care. Of course, this rule is not surprising as it represents an ideally balanced adaptation between the different reproductive species capacities. Which, when reduced, are compensated by a correlative increase in cognitive abilities and adaptability. These, however, come along with an increase in complexity in organisms with lengthening of the gestation of the offspring.

The death, Because?

All the mutations and adaptations of the species aim to increase the number of individuals making up the various gene populations. A large population represents a decisive advantage for the survival of the species and the individual. Thus, all the amazing characteristics of living beings are justified, but why death?

The long-life span of a living organism can be evaluated by natural selection from an adaptive perspective. That feature is very convenient for the species that obtained it since, especially in social animals, a higher number allowed by the longer life span of the individual makes the group of animals stronger. A subsequent level of importance covers the duration of the reproductive life period and the speed of reproduction, which appears to be connected in an inversely proportional manner to the evolutionary superiority of the organism. From the reproductive point of view of the species, the sterile individual does not seem to exist since it does not make any effective contribution to the number of individuals in subsequent generations. However, in the human species, unlike the lower animal species, the best behaviors due to education or the transmission of experience and culture generally lead to a different natural selection, to which the non-reproductive individual also contributes from the biological perspective. For this method of selection, called organic, the ameliorative change in the behavior of individual organisms is equivalent to a selective organic mutation, i.e., a non-genetic improvement that allows more excellent survival skills through a different adaptation of the individual and the community to the different environmental conditions. This particular type of cultural mutation is not necessarily transmitted to subsequent generations as is the chance for genetic characteristics, but allows the species as a whole, through a better and longer time of reproductive capacity, to increase the chances of living and multiply more efficiently.

The life span of different animal species varies greatly. Among the animals with very short life cycles, we find some butterflies that live a few days with an adult phase of a few hours and the butterfly Vanessa of the thistle, whose total life cycle does not extend beyond a year. A little longer is the life span of the Gambusia, a small freshwater fish, in this ranking, followed by mice and hamsters and the variety of Polioptilidae birds, whose lives usually end within the fourth year of life. Man has a maximum life expectancy of 122 years and is among the longest-lived species on Earth. A series of invertebrate organisms greatly exceed these numbers, one of all a Cnidarian, the Hydra, a primitive polypoid animal a few millimeters long, which can remain alive for over fifteen thousand years but potentially immortal thanks to its regenerative properties. If the animal is cut in two, within a few days, its organism completely regenerates, and this process also occurs with minimal parts of its body when dissected. The Hydra can reproduce by budding without the need for sexual reproduction. This phenomenon also means that only environmental conditions endanger its life in terms of pollution and other animals that compete for the same environmental niche but not by some reproductive deficiency.

Man, like other mammals, possesses considerable regenerative abilities both at the cellular and tissue level: it has been calculated that in seven years, all its molecular biological elements are wholly replaced. All the human body cells continually undergo processes of division and proliferation regularly balanced by elimination, which allows proper cell turnover and the complete regeneration of many tissues and organs of the body. A particular example is a liver, which has an excellent capacity for regeneration. There is an almost complete regeneration after some liver damage or surgery (removal of up to three-quarters mass) or viral diseases. This characteristic is due to hepatocytes, the liver cells with an average life of 150 days. Even for removing a part of the liver or after the intake of hepatotoxic substances, cells proliferate and allow complete organ regeneration. Somatic cells in organisms are always perfectly reproduced. This fidelity shows the existence of perfect molecular mechanisms that allow each mother cell to replicate a daughter cell, a perfect copy faithfully.

Among researchers in the field of biology, some questions arise in considering the fidelity of the cells to copy and their replication capacity: if there is a template such as DNA capable of faithfully reproducing itself for an indefinite number of times as, for example, occurs in lines of cultured cell immortalized in culture, so why do animals age and die? Why does a man die? For what strange reason are there biological mechanisms that alter their genetic copying processes and therefore regulate the time for the senescence and death of their organs and tissues?

Indeed, no one dies of old age in the strict sense: there will always be some organ or molecular process malfunctioning. Nobody naturally dies regardless of a specific precision for self-damage inscribed in the genetic code. Old age and death do not seem to follow a central dogma for the Neo-Darwinist theory, which allows anyone to answer any question concerning the reason for the perfection of living organisms and their behaviors. That is the standard answer for which it

The death. Because?

is affirmed that such perfections emerge due to better adaptation of living organisms to the environment and greater efficiency in multiplying the species. Therefore, each organic feature, even the most miraculous, can be explained as an adaptive process of the species to multiply more efficiently and consequently improve its survival. Undoubtedly, old age and death appear against evolution, and they appear as particular characteristics that cannot be seen as a better adaptation than biological evolution imposes on every individual or living species. The explanation proposed for this not very adaptive phenomenon is that if all living beings were immortal, the earth could not be renewed and would soon be insufficient to contain all living organisms. Such an answer would suggest an external evolution plan regarding the living world in general. It is a plan that considers not the individual or species but the whole living world balance and sustainability progress. Therefore, there appears to be a planned suicide in the genetic code for every living organism that is utterly incomprehensible from natural selection's perspective. If we suppose all the perfections of the body are justified by Neo-Darwinist theory, then Why does this hateful feature present itself so entirely contrary to life and the explanatory core of the evolutionary theory itself? This question is entirely legitimate because, in all evidence, death is a characteristic that is not favorable to the species and its expansion, as is, instead, the case for any other organic character.

Man alone, unlike other animals, answers the question by rejecting the idea that one's existence must end. Life has always been an argument religion has brought to support a creator, God evidence. Paradoxically, not life but more death can be adduced as a more than acceptable argument in favor of creationism. Life, at least in its

208

The death. Because?

evolution, can be explained through the concept of the struggle for survival and the best adaptation to the environment, but death cannot. The inexorable perishing of all living beings unequivocally refutes the assertion that all evolution is due to adaptive processes in favor of the species.

So, the significant difficulties encountered in accepting the origin of life starting from the absolute chance focus on two main aspects. The first concerns a beginning from the mineral world, which is difficult to explain from a probabilistic standpoint. While for a second aspect, it remains impossible to understand the beginning of the first molecular mechanisms for the functioning of biological structures, at the same time, capable of both coding for the formation of proteins and programs necessary for their replication conserved in nucleic acids. Furthermore, regarding the evolutionary doctrine, at the current state of our scientific knowledge, death remains incomprehensible because without utility for living organisms, contrary to what is stated for every biological feature emerging in the Darwinian evolutionary framework.

Conclusion

Conclusions

The argumentative scheme of this text shows two theses in comparison and not two proofs. On the other hand, to obtain proof with a value of truth, we must express an affirmation or denial of something and an experience that can still demonstrate the integrity of what has been stated or denied. Aristotle places the apodictic (absolutely certain) syllogism as the only valid and usable one by science: it starts from certain premises to conclude with unquestionable affirmations. I.e., that if "All men are mortal" and "The Greeks are men," then "The Greeks are all mortal." This statement would be a tangible demonstration and invincible if it could be demonstrated by experience, after a tiring census, that the Greeks are all mortal. Still, indeed, the world would stop if one took this type of doubt as the operative basis of acting or thinking.

On the other hand, history teaches those absolute certainties often lead to damage and sometimes even tragedies. Even the Socrates of the first Platonic dialogues never manages to conclude and teach concepts of absolute certainty. With his interrogations, he only moves to doubt the certainties of his interlocutors regarding human and complex things on the type of virtues such as courage, friendship, justice, pity, and so on. Some questions represent pseudo-problems, which cannot be solved since the conditions and initial situations that gave rise to these problems are not verifiable or reproducible. Examples of such problems are those that belong to the mystique of maybe, kind of: "if I had done so, it would have been much better, I would have become rich, I would have had many more friends, and so on. Likewise, the origin of the

Conclusion

living world problem does not find a solution in the experimental certification of the initial phenomena regarding the first form of life.

Regarding the problem of the origin of life, the scientific world is today separated by rigid oppositions in two different factions. It is possible to demonstrate that the two factions have valid logical-mathematical reasons. Robert Hazen writes, "*Given the combinatorial richness of early Earth's chemical and physical environments, events in molecular evolution that are unlikely on limited laboratory scales of space and time may, however, be unavoidable on an Earth-like planet at time scales of a billion years.*"[46].

Tim Barnett on www.str.org/w/building-a-protein-by-chance: "*When you do the math, there are 10^{195} possible ways in which a protein composed of 150 amino acids can be built. The question is, how many of these arrangements are actually functional? Doug Ax of the University of Cambridge has determined that the probability of obtaining a functional protein from all possible total proteins is 1 in 10^{74}.* "These data are aligned with the calculations illustrated in the text in the 'probability and biological molecules' chapter.

However, doubt can be paralyzing, as illustrated by a "sophistic joke" attributed to Gorgia di Lentini in one of his famous writings, "On nature or not being." He said that if something is entirely unknown, it cannot be known because even meeting it, we wouldn't know what it is, and it would therefore not be expressible and cannot be communicated to others. Therefore, such sophisms could trap a logic routed on rigid tracks in our minds. Our imagination, however, must evaluate hypotheses through which we can represent ideas on how something or an even unknown phenomenon can be inserted with logical coherence, at least temporarily, in our cognitive totality, just as a hypothesis. We can see where the unfolding of one hypothesis leads us compared to others and the result, we arrive at can guide us in our actions and lives even on unknown paths.

We got two hypotheses on the origin of life, one concerning creationism by God and the other about the theory of an origin from chance. None can correctly solve that dilemma, and neither of the two theories can be demonstrated scientifically. We can only show the strengths or the weakness of each of them. In the absence of evidence, we are faced with two beliefs or if we want forms of faith that represent free choices that are not equivalent to each other. The implications that involve believing in one hypothesis instead of another for practical action in the world are completely different. The results observed or imagined are the only proof that we can ask of faith or belief.

Regarding the problem of what should be the right answer to problems that a logical-mathematical way cannot solve, we can see the intelligent departures recommended by traffic pundits as an example of a lived life: in particular periods of the year, close to summer holidays or close to other holidays important, they advise to leave at times or days in which, according to these recommendations, we can travel without excessive traffic and delays. Then leaving as recommended is defined as an intelligent way to travel, but how can we be sure that the departure will have been intelligent? We can only know from the result, that is, only on arrival. As sometimes happens during the journey we can encounter hellish traffic and then the departure was stupid and not intelligent. Just as it can happen to a financier who continues to invest money and accepts losses trusting in a final profit: if he ends up ruined,

Conclusion

he was a fool; if he gets rich, he was a genius. Only the result is proof of the goodness of our choices since the initial choice for something we cannot know and foresee is always shrouded in doubt.

In Hegel, in the last paragraph of the Preface of the Outlines of Philosophy of Law, we find the beautiful image of Minerva's owl, which is painted as the bird that begins its flight at dusk. Of course, Hegel refers to the understanding that comes only at the end of the unfolding of thought, history, and obviously also his philosophy. In fact, he writes: *"philosophy always comes too late. As a thought of the world, it appears for the first time in time, after reality has completed its process of formation and is well done. "* [47]

We can anticipate the result deriving from believing in a certain hypothesis over another, through the game of logical implications "if- then ...", so having faith in creationism regarding the origin of life leads to conclusions that are of all different to faith on the origin of life starting from an absolute chance. Only the result of such a premise is a demonstration that can lead to one's own conviction. That is a result with truth value which is, in any case, an own fact. The reason for a specific sense of life lies in the choice.

Bibliography

1) Jacques Monod. Il caso e la necessità (Le hazard et la necessité'1970). ed. 1972, p. 143. Arnoldo Mondadori Publisher Milan.

2) Ibid. page 99

3) Thomas Aquinas: Somma contro i gentili, (Summa Contra Gentiles) UTET. Turin printing-publishing union. Book II, pp. 32-37.

4) Jacques Monod, Ibid. pages 95.96

5) Pier Paolo Viazzo, Introduzione all'Antropologia Storica (Introduction to Historical Anthropology), Bari, Ed. Laterza, 2000, p. 107.

6) C. Diano, L'uomo e l'evento nella tragedia attica (Man and the event in Attic tragedy) (in: 'Dionysus', XXXIX) Ed. Sansoni Editore, Florence (1975)

7) Jacques Monod, Ibid. Page 96

8) Thomas Aquinas, Summa Theologiae, I, question 2, article 3)

9) D.T. Campbell, Epistemologia evoluzionistica (Evolutionary Epistemology). Armando Armando publisher. 1981 pp. 76-77.

10) Erwin Schrödinger: Quantum Theory and Measurement - Part I Section I.11. -J.A. Wheeler and W.H. Zurek, eds., Princeton university Press, New Jersey 1983

11) Hardy G.H., Mendelian proportions in a mixed population, in Science., Vol. 28, n° 706, 1908, pp. 49-50

12) David Reich: 'Who we are and how we got here'. Oxford University Press, ISBN 978-0-19-882125-0, 2018.

13) Arthur Schopenhauer La libertà del volere umano (The freedom of the human will). Bari ed. Laterza 2004, 88-420-7290-7

Bibliography

14) Charles Darwin: : L'espressione delle emozioni nell'uomo e negli animali (The expression of emotions in humans and animals). Paul Ekman Universale Bollati Boringhieri 3rd ed., Turin, ISBN 978-88-33-92296-6.

15) K. C. Lee1, at al., Entangling Macroscopic Diamonds at Room Temperature. Science 02 Dec., 2011: Vol. 334, Issue 6060, pp. 1253-1256

16) BC Hiesmayr, MJA De Dood, W Löffler. Observation of four-photon orbital angular momentum entanglement. Physical review letters 116 (7), 073601, 2016).

17) Juan Yin, Yuan Cao, Yu-Huai Li et al., Satellite-based entanglement distribution over 1200 kilometers, in Science AAAS, vol. 356, n° 6343, pp. 1140 - 1144.).

18) Marletto, C.; Coles, DM; Farrow, T .; Vedral, V. (10 October 2018). "Entanglement between living bacteria and quantized light evidenced by the Rabi cleavage". Journal of Physics: Communications: 2 (10): 101001.18).

19) Kurt Gödel: "Über formal unentscheidbare Sätze der Principia Mathematica und verwandter Systeme", Monatshefte für Mathematik und Physik volume 38, pages173–198 (1931)

20) Colin S. Pittendrigh. Adaptation, natural selection, and behavior in Behavior and Evolution, ed. A. Roe and George Gaylord Simpson, New Haven: Yale University Press, 1958, 390-416; p. 394

21) Richard Dawkins. L'orologiaio cieco: creazione o evoluzione, (The blind watchmaker: creation or evolution). Mondadori libri Spa 2017, pp. 17,39, (En: "The blind watchmaker" 1986)

22) Miller SL. 1953. A Production of Amino Acids under Possible Primitive Earth Conditions. Science 117: 528-529.

Bibliography

23) Luca Bizzocchi et al. Astronomy & Astrophysics. 15 Jun 2020): arXiv: 2006.08401 [astro-ph.GA]

24) Forsythe et al, Proc Natl Acad Sci U S A. 2017 Sep 12; 114 (37): E7652 – E7659

25) Ernst Mayr: The autonomy of biology: the position of biology among the sciences, Quarterly Review of Biology, 1996, 71: 97-106)

26) Gerald Karp, Cellular and Molecular Biology. Concepts and experiments, Edises, 2011, ISBN 978-88-7959-696-1

27) K. Popper, La logica della scoperta scientifica 1934, (The logic of scientific discovery). Einaudi Paperbacks, Turin. 1970 p.154.

28) Hazen RM. 2017 Chance necessity and the origin of life:a physical sciences perspective. Phil. Trans. R. Soc.A375:20160353

29) F. Hoyle, G. Burbidge and J. V. Narlikar, A quasi-steady state cosmological model with creation of matter, in The Astrophysical Journal, vol. 410, 1993, pp. 437–457, '

30) Anton S. Petrov et al - History of ribosome and origin of translation; PNAS 15 December 2015 112 (50) 15396-15401).

31) Jaques Monod, ibid, p. 118.

32) Elizabeth Waters et al: The genome of Nanoarchaeum equitans: Insights into early archaeal evolution and derived parasitism; Proc Natl Acad Sci USA, 28 October 2003; 100 (22): 12984-12988.

33) Allen P. Nutman, Vickie C. Bennett, Clark R. L. Friend, Martin J. Van Kranendonk, and Allan R. Chivas, Rapid emergence of life shown by discovery of 3,700-million-year-old microbial structures, in Nature, vol. 537, 2016, pp. 535–538, DOI: 10.1038 / nature19355, PMID 27580034.

Bibliography

34) Charles Darwin. The foundation of the origin of species. Two essays written. In 1842 and 1844 edited by Francis Darwin. Cambridge 1909 pp. 57-58, translation on John. C. Greene. The death of Adam. Feltrinelli 1971 p. 309)

35) Charles E. Rave. John Ray: his life and works by him, Cambridge University press 1942, pg. 251.

36) (John Ray, The wisdom of God manifested in the works of the creation... 3rd edition London 1701. Preface) on John. C. Greene. La morte di Adamo, (The death of Adam). Feltrinelli 1971 p. 309)

37) William E. Carrol, Creation, Evolution, and Thomas Aquinas, in Revue des Questions Scientifiques, vol. 171, n° 4, Namur, Belgium, Scientific Society of Brussels

38) Allgemeine Naturgeschichte und theorie des himmels, in: Immanuel Kan werke edited by Ernst Cassirer vol. I Berlin. Ediz. Aufbau Verlag Berlin, 1955 p. 44.)

39) K. Popper, La logica della scoperta scientifica 1934, (The logic of scientific discovery). Einaudi Paperbacks, Turin. 1970 p.24.

40) Kimura, Motoo. The neutral theory of molecular evolution. 1983 Cambridge University press.

41) William M. Brandler and coll... Paternally inherited cis-regulator structural variants Paternally are associated with autism. Science 04/202018: Vol. 360, Issue6386, pp. 327-331

42) Junjiu Huang et al. CRISPR / Cas9-mediated gene editing in human tripronuclear zygotes: Protein & cell. May 2015, vol. 6 issue 5, pp 363-372

43) Edward Lanphier et al. Don't edit the human germline. Nature, 12 March 2015, Vol., 519 issues 7544:

Bibliography

44) Julius Fredens, & coll. Nature Article | Published: 15 May 2019.Total synthesis of Escherichia coli with a recoded genome. 569, pages 514–518 (2019).

45) Richard Dawkins, Il gene egoista, (The selfish gene). Oscar essays series edition, Arnoldo Mondadori Editore, 1995, p. 288, ISBN 88-04-) 39318-1)

46) Hazen RM. 2017 Chance necessity and the origin of life:a physical sciences perspective. Phil. Trans. R. Soc.A375:20160353

47) Georg W. F. Hegel: Lineamenti di filosofia del diritto, (Outlines of philosophy of law). translation by Giuliano Marini, Bari, Laterza, 1999, p. 17)

3